305.8
TAT

Tattersall, Ian.

Race?

DATE		

RACE?

NUMBER FIFTEEN: Texas A&M University Anthropology Series

D. Gentry Steele, *General Editor*

Series Advisory Board:

William Irons

Conrad Kottak

James F. O'Connell

Harry J. Shafer

Erik Trinkaus

Michael R. Waters

Patty Jo Watson

A list of titles in this series appears at the back of the book.

RACE?

DEBUNKING
A SCIENTIFIC MYTH

**IAN TATTERSALL
AND ROB DESALLE**

Texas A&M University Press
College Station

This paper meets the requirements of ANSI/NISO Z39.48-1992 (Permanence of Paper).

Binding materials have been chosen for durability.

Library of Congress Cataloging-in-Publication Data

Tattersall, Ian.

Race? : debunking a scientific myth / Ian Tattersall and Rob DeSalle. — 1st ed.

p. cm. — (Texas A & M University anthropology series ; no. 15)

Includes bibliographical references and index.

ISBN-13: 978-1-60344-425-5 (cloth : alk. paper)

ISBN-10: 1-60344-425-4 (cloth : alk. paper)

ISBN-13: 978-1-60344-477-4 (e-book)

ISBN-10: 1-60344-477-7 (e-book)

1. Race. 2. Human evolution. I. DeSalle, Rob. II. Title.

III. Series: Texas A & M University anthropology series ; no. 15.

GN269.T37 2011

305.8—dc22   2011007989

For Bekah, Mia, and Sonya

CONTENTS

ACKNOWLEDGMENTS

AT TEXAS A&M PRESS we would like particularly to thank our editor Mary Lenn Dixon for her enthusiastic support and our copyeditor Mia Scroggs for smoothing out the text. Without the initial encouragement of Ben Roberts, the book would never have been started. Thanks are due also to Jon Marks, Bob Martin, and John Relethford, and to several anonymous reviewers, all of whom made thoughtful suggestions for improving the manuscript, though probably none of these colleagues will entirely approve of the final result. Our gratitude goes also to our wives, Jeanne and Erin, for their forbearance during the entire painful process, and to the American Museum of Natural History for providing an incomparable ambience within which to write. Rob specifically (and very belatedly) thanks his mentors Alan Templeton and the late Allan Wilson, whose teaching, writing, and overall outlook on life have had a huge impact on this book. Many of the more clever passages in this book are the result of remembered lectures from graduate school that Alan Templeton delivered and from the many meetings Rob had with Allan Wilson as a postdoc.

PROLOGUE

ACE. THERE'S probably no more emotive word in the language. The notion it conveys has underwritten some of the worst atrocities ever committed by one people on another, as has recently been happening in Sudan's Darfur province, where dark-skinned "Arab" Muslims are slaughtering and dispossessing culturally and linguistically similar "black" Muslims in what can fairly be characterized as an attempt at genocide. In recent years alone, this is merely one example of many, and throughout recorded human history, perceived "racial" differences have provided the rationale for the infliction of untold amounts of human misery. Clearly, "race" has been a very significant component of our collective human experience. Nonetheless, despite strenuous efforts over two centuries, biologists have been unable to produce any workable classification of the undoubted variety of mankind: clear boundaries among "races" remain highly elusive, as we'll show later in this book.

Races are very evidently not categories, on a par with chairs or trees. What's more, it is becoming ever clearer that those differences among human populations that we intuitively view as racial are not only superficial in terms of import but also of astonishingly recent origin. Among biologists, physical anthropologists are more aware of this fact than most, as indeed they are of the appalling historical record. Painfully mindful of the horrors to which members of their own profession have contributed in the not so distant past, many contemporary physical anthropologists thus prefer to point to the impossibility of classification and to deny that races exist at all. Still, while we will see that there is ample justification for this conclusion, it is hardly a helpful one when anybody walking down the street in a major city can see that "race" does indeed exist in some intuitive sense. The human variety we see around us isn't random, and its order must, at some level, have a biological dimension, since those features that catch the eye are, after all, heritable.

In this book we consider what human races actually are (and are not), and try to place them in the wider perspective of natural diversity. We point out that, for all the agonized exegeses that saturate the literature, the biology underlying racial differences is in fact very simple. First and foremost, we establish that *Homo sapiens* is one single species: one large, interbreeding unit, freely exchanging our genes with each other and with nothing else on the planet. Hardly a difficult thing to do, but everything else follows from this fundamental unity. For within any species, two large-scale genetic processes are possible, each promoting an opposite result. The first of these processes is *divergence*, whereby local populations accumulate genetic and physical novelties that distinguish them from their neighbors. Divergence is typical of all widespread and successful species, and it is most likely to happen where populations are small and scattered, as was the case during the last Ice Age, when our newly evolved species was spreading across the world from an ultimately African point of origin. Through divergence, distinctive "racial" features are acquired in far-flung and isolated populations. The alternative process is *reintegration*, which occurs when differentiated populations that are nonetheless reproductively compatible (i.e., are members of the same species) come back into contact and interbreed. When this happens the boundaries between differentiated local populations become blurred and, if reintegration is given enough time, ultimately disappear.

Contemporary *Homo sapiens* bears the stamp of both processes, reintegration having clearly become the rule once populations of our species had more or less filled up the habitable regions of the Old World. Perhaps ironically, population mixing has become most marked since the adoption of settled lifestyles some ten to twelve thousand years ago, a fateful economic innovation that led to explosive demographic expansion and ultimately to today's unprecedented individual mobility. Yet, for all the diversity we perceive among modern humans, the total timescale involved is incredibly short. From both molecular and fossil evidence we now know that, in *Homo sapiens*, all of the variation we can characterize as "racial" has accumulated in a geological eye blink. We don't currently have any evidence to indicate that our species has been in existence for more than about two hundred thousand years, and all of the variety we see outside of Africa seems to have both accumulated, and started reintegrating, only within the last fifty or sixty thousand years!

The take-home message from all of this is that there is nothing *special* about the racial variation we see within our species: it results from the working of entirely mundane evolutionary processes, such as we encounter in other organisms all around us. It certainly doesn't require any special explanation. In fact, it would be remarkable if we *didn't* see this effect in a species with a history like ours. The short timescale also serves to emphasize just how superficial—and ephemeral—our racial differences are, and, as we point out later, few of them seem to have any major adaptive significance. The reason for this lies in the way in which evolutionary change actually takes place, and this is why, over the course of the book, we will devote considerable space to the process of evolution that underpins our biological history. We will see that, contrary to what many of us were taught, the evolutionary process is not a matter of optimization or of fine-tuning. Chance plays a large role. Where you started from hugely influences where you will end up. And, with luck, just being good enough will get you through. So, without denying that adaptation occurs, we can justifiably point out that many of the differences we observe within species (and indeed among them) are simply random effects, neutral in any functional sense. That they are there means that they are available to be capitalized upon in the future, should the occasion arise; for the nonce, however, they don't mean much if anything at all.

In another irony, random innovations may in fact be more accurately descriptive of particular geographical populations ("races") than adaptive ones are, since fewer forces favor them and they are thus less likely to be independently acquired. To illustrate this we might cite skin color—which is one of the few varying human traits that we *can* truly describe as adaptive, since dark pigmentation is vitally protective in the tropics, yet may in some respects be a liability at high latitudes. And in this case, adaptation turns out to be a very poor guide indeed to history. Bantus, native Australians, and Tamils are all typically very dark-skinned but have entirely different geographical and temporal histories—as do light-skinned Europeans and northern Asians. The result is that, whatever those entities may actually be that we habitually describe in the United States as "blacks" and "whites," they lie in the realm of sociocultural constructs rather than in that of biologically coherent groups.

Intellectual constructs of this kind may, of course, involve aspects of biology. But, much more importantly, they are intricate compounds of historical, economic, and cultural elements, and it is beyond either our scope or our competence to tread on that terrain here. Indeed, we believe that biologists should intrude into such areas only with the greatest caution. Biologists are, of course, uniquely qualified to draw attention to the routine nature of the evolutionary and genetic processes that have underwritten local physical differentiation in *Homo sapiens*. Indeed, emphasizing this simple fact is without question the most valuable single contribution they can make to the race debate. There is also no doubt that biology alone can unravel the history of humankind as an actor in the great evolutionary play and explain, on the physiological level, how it was that our species came to conquer every inhabitable zone on Earth. To members of an egotistic species, "human biology" is inevitably one of the more fascinating areas of science, and this is why we will explore it in later pages. But the issue of "race" goes far beyond strict biology, to encompass many of the more murky and unfathomable aspects of the human psyche. The problem in Darfur, as elsewhere, is at heart one of clashing historical, cultural, political, economic, and social identities—and not one of biology at all.

For millions of years, human beings and their precursors typically lived in small social groups confined to relatively tiny territories. Repeated studies have shown that, even today, it is difficult for most individuals to identify closely with more than about 150 others. Although it is possible to maintain some level of group identity among much larger numbers of people than that (using elaborate coercive mechanisms), beyond a certain point it seems somehow built into the human psyche (as it is among primates in general) to begin regarding individuals outside the immediate social unit as "other." Of course, various degrees of "otherness" are commonly recognized. But history shows that the perception of otherness is often amplified in the presence of physical distinctions of the kind usually regarded as racial, sometimes to the extent of denying the humanity of the other. Recent scientific explanations of such behavioral phenomena often rely on sociobiological arguments, blaming this or that ugly behavior on lingering adaptation to conditions that prevailed in some remote and probably entirely mythical past. Accounts of this kind, often cast within the discipline of evolutionary psychology, appeal greatly to our reductionist minds, but on close inspection their rigor often turns out to be questionable.

One area in which many have seen hope for improving the human lot by invoking racial classification is medicine. Many clinical conditions are associated with particular gene configurations and have been reported at unusually high frequencies in particular groups of people. This has resulted in what have been called "race-based genomics" and "race-based medicine," as means of discovering and coping with disease pathways in our species. However, as we will explain in our discussion of the human genome, there is a whole host of analytical problems with this approach. Furthermore, it is not populations (however they may be defined) that fall ill, but individuals, and we will see that belonging to any socioculturally defined group is usually a remarkably poor predictor of exactly which genes you will have. Now that it is becoming clear that individualized genomes will be economically feasible in the near future, we expect that the clinical emphasis will almost certainly move in that direction.

Genomic studies have also provided us with a new and unprecedented way of tracking how modern humans took over the planet following their amazingly recent exodus from the continent of their birth. We will see in our discussion of the human fossil record that, while fossils are quite abundant for earlier periods of human prehistory, they are less useful when it comes to the very compressed timeframe of modern human emergence. Fortunately, once our species is actually on the scene, tracking the histories of genetic markers (which can be done via comparative analyses of modern populations) provides us with by far the best way of following the initial expansion of humans out of Africa, and ultimately into the farthest reaches of the inhabitable globe. At some point (we're not there yet) we will have enough information to see how this history of spread maps on to the history of physical differentiation in *Homo sapiens*. Yet population studies already also bear eloquent witness to the extraordinary levels of population intermingling that have occurred in subsequent phases of this eventful history.

This book is, then, mainly about the *limits* of what biology can usefully say about the nature and existence of human races. New techniques and new approaches can and will tell us an enormous amount about the biological history of our species, but they also teach us that this history was a very complex one that is very inaccurately—indeed, distortingly—summed up by *any* attempt to classify human variety on the basis of discrete races. While we can acknowledge that our ideas of race do in some sense reflect a historical reality, and that human variety does indeed have biological underpinnings, it is important to realize that those biological foundations are both transitory and epiphenomenal. Despite cultural barriers that uniquely help slow the process down in our species, the reintegration of *Homo sapiens* is proceeding apace. And this places the notion of "races" as anything other than sociocultural constructs ever more at odds with reality. Increasingly, it seems, we are simply who we think we are.

RACE?

CHAPTER 1

RACE IN WESTERN SCIENTIFIC HISTORY

AS A CULTURAL, historical, and political phenomenon, race still permeates modern society in a way that nothing else does. The issues it raises are complex, and, to make the situation more difficult, they are deeply embedded in the often opaque human psyche. For scholars this makes the intricate linkages between the varied aspects of the race question extremely difficult to disentangle, just as in individual minds it is hard to disengage the intellectual from the emotional components of the issue. In a species as cognitively complicated as ours, things could hardly be otherwise. Still, we do believe that the social conversation would be vastly simplified—and improved—by subtracting many if not most aspects of biology from it. The reasons for this belief are numerous, as we will see, but one place to start making the point is that the historical record of "science" in this arena has been very far from stellar. As only the most obvious example, notions propounded by scientists about the innate biological "inferiority" or "superiority" of particular races have underwritten some of the most atrocious socioeconomic policies of the past two or three centuries—without actually possessing any underpinnings in objective science.

Sadly, though, with a few notable exceptions, scientists over the years have done little to counter the notion that race in the larger sense and biology are inextricably entangled, and some have actually fanned the flames. As a result, there is still a huge amount of misunderstanding about what "race" really is in biological terms, and its intrinsic importance and functional significance to larger society have consequently been enormously overestimated. We have written this book because we see an urgent need to put "race" into appropriate perspective and to look at just what science can and cannot usefully say about this sociologically and historically important issue. Our conclusions will be very simple; indeed, many will find them anticlimactic. First, we will point out that, as far as evolutionary biology is concerned, race is a pretty routine matter: so unremarkable, indeed, as to require no special explanation from biologists. Second, we strongly believe that emphasizing this simple reality is the single most important contribution that scientists can make in the arena of race. Indeed, it may be the *only* one that, as dispassionate observers, they can really helpfully make on the subject.

Human populations undeniably differ in various visible characteristics. Biology has shown such variation to be typically trivial (in biological terms, selectively neutral), and always epiphenomenal; yet it is equally true that hideous historical excesses have been committed in the name of the differences perceived. Given this unfortunate combination of realities, it appears imperative to many people that science and scientists both cannot, and should not, stand apart from the larger social debates about race, especially where inequity is implicated. But at the same time it must be recognized that the record of science as an arbiter of social policies is hardly a distinguished one, and it is equally true that, merely to preserve its unique identity, science is obliged to maintain a distance from the sociopolitical arena. Mixing science with politics has all too often resulted in such destructive distortions as the imposition of Lysenkoism upon the nascent science of genetics in interwar Russia, or of *Rassenhygiene* (racial hygiene) on German and other European societies in the 1930s and 1940s. Science can, of course, often provide information that is useful to society in making decisions about what is moral or desirable (or not). But it cannot itself make those decisions. So, while it seems self-evident to many that science should be socially engaged and ought to be placed specifically in the service of what in the light of scientific or other knowledge seems right, it's just not that simple. Science has been hugely abused in the past in the name of sociopolitical agendas, and if scientists are going to have any sociopolitical agenda at all, it is surely to prevent this from happening again.

History, then, suggests that beyond the flow of essential information from science to society, extreme caution is in order. For scientific knowledge is truly *different* from any other way of knowing, most importantly in being inherently *provisional*. Scientists are not seeking enduring truths; they are simply edging in upon a fuller and more accurate description of the world around them, by proposing testable new ideas about it and discarding those that are shown to be wrong. Composed as it is of falsifiable bricks, the scientific edifice will never be completed; indeed, even its very foundations are theoretically at risk from advances in testable knowledge. In stark contrast, moral and ethical judgments are at least in principle absolutes—even though our views of what is just and right may change. And in many ways politics exists either to exploit or to implement the imperatives that emerge from those absolutes.

We have said that our main aim here is to point out the limits of what biology can say about race. And the implication is, of course, that the rest is up to the social scientists, philosophers, and politicians. This we truly believe. But some historical background to the issue of race and science nonetheless seems necessary, if only because scientists have historically been so ready to exceed their remit and to pontificate on aspects of race that carry strong political and historical and social implications. This is why we're starting off this book with a chapter on the history of race in western thought, the intellectual tradition from which modern science emerged. We don't claim that this chapter provides anything new, either in fact or in conceptual framework, and we have certainly not tried to be comprehensive, even within the Western canon. The subject is far too vast. We also recognize that, viewed from other perspectives, the sociopolitical implications of this history may well appear significantly different than they do from within science. But this book is written for an audience that is, for better or for worse, deeply influenced by the Western tradition that gave birth to modern science, and we think it important to sketch here the main lines of developing thought about race in the western world—as a way of explaining the devious path by which biology has reached both its modern perspectives and its current predicament.

EARLY INTIMATIONS OF HUMAN DIVERSITY

We human beings are many things. But one attribute we all seem to have in common is an urge to classify everything around us. That's the way our brains work, and it is not coincidentally the way we communicate, too, naming and thus in some sense pigeonholing virtually everything that can be objectified in any way. After all, this is something we have to do if we want to talk or even to think about those things. So it's not surprising that we are instantly ready to classify and categorize each other as well. We are exquisitely sensitive to the variations we perceive among the people around us, whether in dress, class, economic status, appearance, accent, or language, and we respond to these variations by an instinctive categorization: He is dressed like a banker; she is speaking German; someone else looks Asian.

Imagine yourself, then, in Europe as it slowly emerged from the Middle Ages. Nothing could have been further removed from life in today's great cities, where the streets teem with an almost unimaginable diversity of people. As the Age of Reason dawned, most Europeans' life horizons still hovered within a few miles of where they were born, and almost everybody with whom an individual might have interacted on a daily basis was a family member, a neighbor, or at most from the next village or town. Societies were small and homogeneous; and even in great capital cities such as Paris or London, small towns by modern standards, the occasional foreigner was most likely a European member of a tiny international noble or royal class or maybe a prosperous merchant. The key was homogeneity, and class and occupation divisions within European societies were vastly greater than the overall differences among them. Beneath their often extravagant and symbolic clothing almost everyone, even one's hated neighbor, *looked* basically the same. For those with access to classical literature there were, of course, ancient tales of barbaric tribes and exotic peoples, dating back as far as the days when the armies of Alexander the Great were cutting a swath into the underbelly of central Asia. And to those on the exposed eastern flank of Europe, the thirteenth-century Mongol invasions were still fresh in popular memory. But in most of Europe, where the "scientific" study of race began, such semi-mythical peoples often ranked alongside the fantastical denizens of bestiaries, with little if any relevance to daily life other than as characters in a morality tale.

What, then, were European intellectuals and others to make of the stories—and the occasional exotic individual—brought back by early explorers as the Age of Discovery began in the fifteenth and sixteenth centuries? Our species is highly intolerant of boredom; and at a time when much public amusement in the cities was furnished by such pastimes as bear-baiting, cock-fighting, the stocks, and public executions, the occasional arrival of exotic-looking people caused huge excitement, and stereotypes rapidly emerged of the peoples at the far ends of the rapidly developing trade routes. The public lapped up such accounts as those of Antonio Pigafetta, one of the only eighteen survivors of the epic Magellan round-the-world expedition of 1519 to 1522, who claimed that a native of Tierra del Fuego at South America's southern tip was so tall that "we reached only to his waist." Later explorers over the next two centuries were more than happy to embroider on the idea of "Fuegian giants," even though the Fuegians themselves, when a group of them finally arrived in London in 1830, proved to be disappointingly modest in height.

Unlike those first Fuegians to visit England, who arrived as something in between the guests and captives of Robert Fitzroy, captain of the round-the-world voyage of the *Beagle* during which Charles Darwin first formulated his evolutionary ideas, the first Africans to come to England had arrived more or less as slaves. They were brought from West Africa in 1554, by the notorious Capt. John Lok, to assist in the development of the burgeoning slave trade. Actually, while they were probably slaves on the ship, they were "free" once they stepped ashore, for slavery had no statutory recognition in England, and in any event they eventually returned home, to unrecorded fates. Still, slaves or not, these were hardly the most auspicious of circumstances in which to reach a new country, and indeed, most of the early unfamiliar faces to come to Europe did so in disadvantaged conditions, as slaves or at best as curiosities.

Familiarity rapidly bred unease, perhaps even fear, and as early as 1596 and again in 1601, Queen Elizabeth I, deploring African "infidels" as freeloaders, issued orders expelling them from England. The ploy failed because not a few Africans had actually managed quite successfully to integrate themselves into English society, but clearly the expatriates had acquired an image problem in some quarters at least. By the beginning of the seventeenth century, then, social attitudes towards exotic people in the midst of European society were already showing signs of being as conflicted as they were to become in later times. As with almost all social attitudes, this was nothing new. Among other ancient societies the Greeks, Romans, and Egyptians had all classified the peoples they encountered in outlying areas of their known worlds (and not a few mythical and monstrous ones as well), using much the same features as we do today, and imputing various disreputable character traits to them. In a more scientific spirit of inquiry, from the ancient Greeks onwards a tradition was established of attributing physical differences among peoples from different parts of the world to environmental influences, to different "Airs, Waters, and Places," as Hippocrates put it.

BIBLICAL INFLUENCE

The first intellectual debates in Europe over the status of exotic peoples took place in the context of Christian theology. As interpreted by the medieval Scholastics, defenders of orthodoxy, the Bible furnished the accepted account of human origins. According to the first ten chapters of Genesis, all human beings were descended from Adam and Eve, and all forms of mankind from Noah's curious variety of sons. Everyone was, as it were, in the family, albeit a family with three branches. The default position was thus the "monogenist" one: a single origin for all of humankind, diverse as it might be.

Yet by the sixteenth century this notion was beginning—very occasionally—to be questioned by heterodox thinkers. In 1520 the humanist physician Paracelsus had the temerity to suggest that people living in far-off places "did not descend from Adam." And around a century later the Italian philosopher Giulio Cesare Vanini, perhaps tongue-in-cheek, raised the possibility that humans were in some way related to or even descended from monkeys. Sadly, this suggestion got him strangled and burned at the stake, in the central square of Toulouse, by a Catholic church that was not renowned for its sense of humor. Half a century later yet, the Calvinist Isaac de la Peyrère saw not himself but his books publicly burned in Paris for suggesting, among other things, that "if Adam sinned . . . there must have been an Adamic law according to which he sinned. . . . If law began with Adam, there must have been a lawless world before Adam, containing people." The clear implication of de la Peyrère's "Pre-Adamite" suggestion was of multiple creations, and while this heretical notion did not go down well at the time with the Catholic authorities, it sowed the seeds of energetic later debate as the monogenists took issue with the "polygenists" who favored separate origins for the races.

In all of this it has to be borne in mind that there were practical as well as theological considerations at issue for the Catholic Church. Much of the Iberian conquest of South America in the sixteenth century was carried out under the veneer of a Christianizing mission. This meant that, where they could not be hidden or denied, the horrible abuses inflicted on natives by the conquistadors had to be somehow justified. One avenue for doing this lay in the polygenist argument: if the Amerindians were a separate creation, then they had essentially the same status as beasts of burden, and no rights as human beings. To his credit, Pope Paul III took the high road on this one: in 1537 he issued a Papal Bull declaring that the Indians were "true men . . . capable of receiving the Christian faith . . . [who] must not be deprived of their freedom and the ownership of their property." Sadly, this did little to stop the abuses, and things got so bad that thirteen years later, the Habsburg ruler of Spain convened a conference on whether the colonization process should be continued. Clerics made impassioned arguments on both sides, and, largely due to eloquent pleading of the cause that "all men are human" by the Dominican friar Bartolomé de las Casas, colonization was eventually legally suspended. The respite was, alas, temporary, as commercial and political interests reasserted themselves over moral ones.

THE ENLIGHTENMENT

On the more secular front, and further from the political fray, the intellectual landscape was rather different. As early as the late fifteenth century the first reports of Columbus (who had initially headed west in high trepidation at the thought of encountering giants and assorted monsters of the kind described by Pliny and others) had begun to usher in the idea of a sort of New World Arcadia in which people lived in harmony with Nature and without the constraints of European society; in the seventeenth century this dreamy illusion was developed by Enlightenment philosophers into the ideal of the "noble savage." This was the iconic image of pristine peoples who had existed in Europe in the remote past and who still occupied far-flung lands. They lived in harmony with nature and without need of what then passed in Europe for creature comforts. The image was a seductive one, harking back as it did to an earlier Eden for which Europeans living in the crowded, noisome cities of the sixteenth and seventeenth centuries yearned. In this spirit, the failure of New World peoples to invent ironworking technologies was taken less as a matter of technological inferiority than as an indication of a kind of simplicity and purity that had been lost in European society.

Imperfect as it was, the burgeoning awareness of the diversity among human beings around the world in appearance, social structures, and lifestyles began as time passed to have profound effects on progressive philosophers. It both provoked and underwrote the search for what the Scottish historian and moral philosopher David Hume described as "the constant and universal principles of human nature." So convinced was he of these principles that in 1748 Hume declared that "Mankind are so much the same, in all times and places, that history informs us of nothing new or strange in this particular." Not that everyone agreed. Hume's friend Henry Home, Lord Kames, another leader of the Scottish Enlightenment, is celebrated as the judge who had ruled in a famous case that there could be no slavery in Scotland. Nonetheless, in a remarkable book entitled *Sketches of the History of Man*, Kames argued in 1776 that the differences among human races were so great that climate, the factor to which they were still customarily attributed, could not in fact have caused them. This hinted, of course, at a polygenist explanation for human variety. And even while most Enlightenment philosophers sided with Hume rather than with Kames on the essential unity of mankind, opinions were still divided on what the common ingredient was. Shortly after Hume wrote in defense of basic human decency, Thomas Hobbes in England decided that the basic human state everywhere was one of continual conflict. The "life of man" was "solitary, poore, nasty, brutish and short," so that social constrictions were the only glue that held complex society together and permitted civil interaction. Across the Channel views were equally varied. Jean-Jacques Rousseau, the most renowned proponent of the noble savage, rejoined that "we shall not conclude with Hobbes that just because he has no idea of goodness, man must be naturally wicked." While Voltaire (probably mischievously, but unfortunately nonetheless) declared that the human races looked and behaved so differently that they must surely be different species.

Of course, in the eighteenth century the concept of the "species" was exceedingly poorly defined—as in some respects it still is. Species were the various "kinds" of organisms out there, but exactly how they were to be recognized was unsure. Today most biologists would agree with the seventeenth-century English anatomist John Ray that they are essentially reproductive entities—the largest groups within which unrestricted interbreeding occurs, or is possible. This is not by any means an easy criterion to apply, and we know now that over time changes may occur. But three hundred years ago, while it was accepted that species had been created more or less as they were, how they were bounded was the grayest of areas, with the result that terms such as "species," "race," and "variety" were often used more or less interchangeably—and continued to be for a long time.

As the eighteenth century began, not all philosophy was as forward-looking as Hume's, and science itself was in its infancy. Indeed, the boundary between philosophy and science was all but invisible, a conflation that is still ceremonially preserved in some quarters: not enormously long ago, one of us attended a medieval university that still offered degrees in "moral sciences" to philosophers and in "natural philosophy" to scientists. A huge tension still existed between the view of the natural world inherited from Classical sources such as Aristotle and the emerging outlook of the Enlightenment philosophers epitomized by Hume, who considered that reason and observation were the only legitimate bases for knowledge.

One of the most durable concepts stemming from the Classical canon was that of the Great Chain of Being. Hugely elaborated by the Scholastics in medieval times, this was a notion that saw everything in the universe as ordered in a hierarchy that started with rocks and minerals at the bottom and proceeded inexorably upward, through plants and "lower" and "higher" animals, to human beings. These exalted creatures sat just below the angels, and God presided at the top. Every species and variety of living organism had its place in this hierarchy; and in the eighteenth century and later, the Great Chain notion underpinned various versions of the belief that human races could be similarly ranked from "lower" to "higher."

HUMANITY'S PLACE IN NATURE

Of course, Europeans at the time hardly had a very complete view of the diversity of humankind or even of the natural world in general. This was particularly true of humankind's immediate zoological context, and even at the very end of the seventeenth century European science still knew very little about the existence in far-flung places of human-like primates. The famous Dutch anatomist Nikolaas Tulp published an anatomical illustration of a chimpanzee in 1641, but substantial inquiry into the comparative anatomy of primates only really began in 1699, when the English anatomist Edward Tyson published his *Orang-Outang sive Homo Silvestris: or, The Anatomy of a Pygmie Compared with that of a Monkey, an Ape, and a Man*. The welter of references in Tyson's title to the "Orang-Utan," "Homo Silvestris" (man of the woods), the "Pygmie," and other forms show just how fragmentary knowledge of the living world was at the time. Nonetheless, in this seminal work Tyson showed an amazingly modern turn of mind and did exactly what the latter part of his subtitle suggested, comparing the anatomy of a juvenile chimpanzee from Angola with that of a modern human and a monkey. He demonstrated that the ape resembled humans in forty-seven key respects, while resembling the monkey in only thirty-four. From this he concluded that the chimpanzee was neither a kind of human nor a variety of monkey but a separate form that lay intermediate between the two. What's more, he deduced that the "pygmies" and "satyrs" described by Classical geographers cannot have been bizarre forms of human but instead were either apes or monkeys. This sober reasoning provided a firm and remarkably prescient platform for positioning humankind within a scientific view of nature. But it could also, of course, still be interpreted within the Great Chain of Being.

The outlook embodied in the Great Chain later came back to haunt "scientific" analyses of human variation but not for a remarkably long time. Indeed, the very first

organized consideration of the geographical varieties of mankind did not hew in the
least to this natural hierarchy. Instead, rooted in observation and personal experience,
it lay firmly in the nascent Enlightenment tradition. Published in 1684 under the title
of *A New Division of the Earth by the Different Species or Races which Inhabit It,* it was
the work of a French physician, François Bernier. This remarkable man had travelled
widely in the Indian subcontinent and knew intimately the complexities of several
non-European societies. Bernier recognized Europeans, Africans, Chinese/Japanese,
and Laplanders as "races of men among which the difference is so conspicuous that it
can properly be used to mark a distinction." But at the same time, over half a century
before Hume, Bernier declared that all these races were differentiated by little more
than the color of their skins. Psychologically, all were one.

An auspicious start, but once a formal framework had been proposed in the
mid-eighteenth century for classifying the denizens of the natural world, it was almost
inevitable that the varieties of humankind would find themselves being classified in
the same way. The modern system of classifying living things was developed by the
eighteenth-century Swedish botanist Karl Linné, who wrote (in Latin) under the
Latinized version of his name, Linnaeus. Published in 1758, the tenth edition of his
great work *Systema Naturae* is taken as definitive. Linnaeus assigned each organism to
a species that was identified by two names, for example *Homo sapiens.* The first name
(Homo) denoted the genus to which the species was assigned; the unique combination
of the two names denoted the species. Just as each species was assigned to a genus,
every genus was assigned to an Order, each Order to a Class, and each Class to a
Kingdom. In this way Linnaeus created an inclusive natural hierarchy of nested sets, in
which every "taxon" (the name given to a group at any rank level) belonged to each of
the categories above it. The genus *Homo* lay in the Order Primates, which belonged in
the Class Mammalia, and so on. This was indeed a hierarchical arrangement; but the
Linnaean system nonetheless contrasted fundamentally with the linear "military" style
of hierarchy exemplified by the Great Chain of Being, in which each kind of organism
occupied a sole and unique position.

Despite this fundamental difference in philosophical approach from that of the
Scholastics who had elaborated the Great Chain of Being in the service of the biblical
account of Creation, Linnaeus had no problem in reconciling this new way of looking
at things with the fairly literalist Christian faith to which he and most members of

European society subscribed. Linnaeus did not go uncriticized, but as long as the idea of the fixity of species was not attacked, and as long as the pattern of resemblances among organisms on which Linnaeus had based his ordering of nature was seen as simply representing the will of the Creator, most people saw no threat to the fundamental beliefs of the day. Indeed, to Linnaeus, never one to underplay his own achievement, he and God formed a sort of team: "God created; Linnaeus classified."

Even so, it was a bold move for Linnaeus to classify human beings as just one more product of nature. Even bolder was his classification of chimpanzees along with our species in the genus *Homo,* as *Homo troglodytes.* What this assignment also expressed, however, was the uncertainty that still surrounded the status of the poorly known primates that most closely resembled humans. Very little was known in any systematic way, and this lack of context was reflected even in Linnaeus's diagnosis of the genus *Homo.* One of the most important innovations in the Linnaean system of classification was the use of a few unique physical characters to differentiate each taxon, which was then grouped with others on the basis of shared similarities. Modern humans have more than their fair share of physical peculiarities that mark them off from any other living form, yet in the case of *Homo* Linnaeus curiously contented himself with the admonition "nosce te ipsum" ("know thyself"; as we will see in the next chapter, Linnaeus's propensity for knowing himself led to his being named the type specimen for *Homo sapiens*).

Still, just as Linnaeus saw similarities between chimpanzees and people that were compelling enough for him to classify them together, he also perceived enough differences among human populations to depart from his normal practice in classifying animals and to recognize taxa *within* the species *Homo sapiens.* He did not name the infraspecific taxon at issue, as he had named the ranks of species, genera, and so forth, but it seems reasonable enough to compare it with the units variously described today as "subspecies" or "races" or "varieties." His first human infraspecific form was "*Ferus,*" (wild), a four-footed, mute, and hairy creature that apparently corresponded to the folktales abounding in his days of wild children raised by animals. His last was "*Monstrosus,*" which referred to a collection of mythical or quasi-mythical humans that included those Fuegian giants. But in between he listed four readily recognizable regional types that he recognized principally on geography and skin color and characterized in terms of the four basic types of temperament recognized by the physician Galen in the early eleventh century (a classification that was itself based on the four "humors" first proposed by Hippocrates in the fourth century BC). First in the list was *Americanus*, the "choleric" Amerindian group with red skins. These people, said Linnaeus, were energetic, upright, and combative. Next came the "sanguine" *Europaeus*: white, confident, muscular, and inventive. Then followed the "melancholic" *Asiaticus*: yellow, gloomy, thoughtful, inflexible, and avaricious. Finally came the "bilious" *Afer*: African, black, self-contented, lazy, slow, and relaxed. Throughout, Linnaeus's characterizations were consciously idealized. As the anthropologist Jon Marks recently pointed out, Linnaeus described Europeans "as having blue eyes and blond, flowing long hair . . . while knowing full well that the vast majority of Europeans possessed neither of those features. His purpose was not empirical, but ideal. . . . he was describing what [Europeans were] *supposed* to look like."

This, then, was the first formal attempt to classify human beings alongside the rest of nature. It was clearly an essentialist one, and in departing in several respects from Linnaeus's treatment of other animals, it already showed at least a minor degree of human exceptionalism. Right at the beginning, then, it ushered in the kind of stereotyping that inevitably occurs when anyone tries to shoehorn such a locally diverse species as *Homo sapiens* into neat geographical categories.

A more fleshed-out consideration of human geographical variation emerged at about the same time from the pen of the French polymath Georges-Louis Leclerc, Comte de Buffon, thirty-five volumes of whose widely read great work *Natural History: General and Particular* appeared between 1749 and his death in 1788 (more were yet to come, edited by his former assistant, Bernard-Germain Lacépède). Buffon was a remarkable intellect and in many ways a freethinker. He claimed to be a Christian believer, yet he accepted that all phenomena had natural causes, denied any belief in Noah's flood, and dismissed biblical estimates of the age of the Earth as far too short. He speculated about the origin of life on Earth and even entertained the notion that humans and apes, so structurally similar, might share a common ancestry. His study of vestigial organs in a variety of vertebrates led him to the conclusion that organisms could leave old functions behind them and even acquire new ones. Yet he refused to believe that one species could change into another. Indeed, in 1753, over a hundred years before it was formally proposed by Charles Darwin and Alfred Russel Wallace, he specifically rejected the as-yet-unarticulated notion of evolution: "If it were once proved that [nested sets] could be established rationally . . . there would be no limits to the power of Nature. . . . [S]he could have drawn with time, all other organized beings from a single being. . . . But it is certain, from Revelation, that all . . . species emerged fully formed from the hands of the Creator." Buffon's eventual resort here to Christian doctrine seems very much like a convenience, a literal *deus ex machina*, as though it were just too much (or too dangerously) in contradiction of received wisdom to follow his initial line of reasoning through to its logical conclusion. But it also fits with Buffon's remarkably modern view, elaborated from John Ray's earlier notions, that species are reproductive entities with self-defining boundaries. And it was probably because of his belief in the impermeability of those boundaries that Buffon had no interest in classifying species into larger groups. To him the species was the basic unit of nature, and there were no larger ones. We'll come back to a detailed consideration of what species are, and how geneticists and systematists approach them today, in the next chapter of this book.

All we know of Linnaeus's views on human races in general is what we can glean from his telegraphic classification, in which *Homo* was simply ranked alongside all the other living things that Linnaeus was acquainted with. Buffon, in contrast, surveyed the rapidly accumulating literature of travelers' tales, already quite extensive by the second half of the eighteenth century, and rather casually used the term "race" to denote current local varieties of humankind. He was impressed by the cultural and temperamental uniformity he perceived among the inhabitants of the New World and concluded "independent of theological considerations" that their origin was "the same as ours." He even remarked on the similarity of the Amerindians to the natives of Kamchatka and suggested that the Americas had been colonized from Siberia. Moving on to the continent of Africa, Buffon noted in contrast the "variety of men it contains" and concluded that this variety (principally in skin color) was "very ancient" and must have been caused by climatic factors, as elsewhere in the Old World. "On the whole," he concluded, "mankind is not composed of species essentially different from each other. . . . [O]n the contrary, there was originally but one species which, after multiplying and spreading over the whole surface of the Earth, has undergone various changes by the influence of climate, food, mode of living, epidemic diseases, and the mixture of dissimilar individuals. . . . [A]t first these changes were not so conspicuous. . . . [They] became afterwards . . . more strongly marked . . . and will gradually disappear . . . if the causes that produced them should cease." Buffon may not have been an evolutionist, but his view of the varieties of mankind as dynamic, responsive, changing entities has a breathtakingly modern ring to it.

His words certainly had an impact on the immensely influential philosopher Immanuel Kant. In the strict realm of philosophy, Kant's major objective was to bridge the widening divide between the empiricists who believed that knowledge is gained solely through experience and the rationalists who maintained that knowledge comes from reason alone. When he strayed into biology, though, Kant showed distinctly rationalist sympathies. Allowing for some differences in terminology, Kant was impressed by Buffon's genealogical notion of the species, and he grouped all the races of mankind together because they could hybridize. But he was much less interested than the empiricist Buffon in the processes going on inside this species. Writing in 1775, Kant concluded (like Linnaeus) that there were basically four races of humans. As Kant saw it, these were northern Europeans (blond, adapted to cold and damp), Americans (copper-red, adapted to dry cold), Africans (black, adapted to dry heat), and Indians (olive-yellow, also adapted to dry heat). Kant correctly pointed out that in similar environmental conditions you did not necessarily find the same race. And from this he went on to derive a conclusion that was the polar opposite of Buffon's: once a particular form or physiognomy was in place, he claimed, there was no changing it. Only the "stem" form was susceptible to modification, and once the derivative forms were established, there was no going back. This was a philosopher's conclusion, not a biologist's, but Kant's essentialist notion made a profound impression that many biologists found it hard to shake off.

NOTIONS OF CHANGE

Another who departed from Buffon's way of viewing things was Jean-Baptiste Pierre Antoine de Monet, Chevalier de Lamarck, one of his successors at the Jardin des Plantes in Paris. Around the turn of the nineteenth century, Lamarck's studies of fossil mollusks led him to the conclusion that lineages of animals had indeed changed over time. Via an innate "life force," organisms showed a tendency to increasing complexity, and adaptation to local environments was achieved via the use and disuse of physical structures. Lamarck's was the first-ever coherent statement of an evolutionary process, but alas it became discredited in the eyes of history by an unfortunate choice of mechanism: the notion (albeit one widely accepted at the time) that characteristics acquired during the lifetime of an animal could be passed along to its offspring. It was particularly unfortunate for Lamarck that his ideas of slow change were immediately and savagely attacked by his enormously influential contemporary Georges Cuvier—who long outlived him, and whose eulogy at Lamarck's memorial service was widely if somewhat unfairly taken to heap ridicule on his ideas.

Cuvier was a paleontologist of great distinction. He was also the leading advocate of "catastrophism" as an explanation for the changes he saw over time in fossil faunas that he saw were mostly composed of animals now extinct. To reconcile this undeniable fact of change with Christian beliefs that included the fixity of species, Cuvier propounded the notion that extinct forms were the remains of animals swept away in the biblical flood. Realizing that a single flood was incompatible with the succession of extinct faunas he observed in the geological column, he eventually accounted for faunal change by invoking an entire long series of ancient catastrophes, each followed by immigrations or new creations. Clearly, between the Lamarckian and Cuvierian positions there was no room for compromise.

In terms of evidence, Lamarck and Cuvier were actually looking at different sides of the same coin (there is evidence in the fossil record both for transformation and for extinction), and it is significant that, even with such radically different views on the origin of diversity in the living world, they converged on the matter of the essential unity of mankind. Lamarck's evolutionary views led him to propose in 1809 that modern bipedal humans had descended from a rather ape-like quadrupedal ancestor and that this gradual transformation had allowed them to impose "an absolute supremacy over all the rest." Once physical superiority had been achieved, the acquisition of modern human cognitive faculties followed naturally, via adaptation, as needs gave rise to efforts that in turn gave rise to modified structures and abilities. Lamarck was not specific about how the current diversity of humankind had come about, but clearly he shared the view that it was through accommodation to environmental influences, and that racial differences were merely variations on a theme. Probably following the German scholar J. F. Blumenbach, of whom we will hear more, he recognized six human "varieties": the Caucasian (European), the Hyperborean (Eskimos), Mongolian (Central Asian), American (Amerindian), Malayan (South Asian), and Ethiopian (African). However, even this recognition was essentially a pro forma one since, being more preoccupied with changing lineages than with static taxa, Lamarck had little interest in classification.

The same can probably also be said for Cuvier, whose focus was much more upon the detailed understanding of anatomical structures than on classifying the entities that contained them. Like most scientists of his day he accepted that species could vary within themselves but regarded the barriers between them as absolute. From the beginning, impressed by the "promiscuous intercourse" among humans of all geographical origins, Cuvier saw human differences as those among members of the same species; and while early in his career he inclined to the view that the physical differences associated with race extended to intellectual potential as well, his stance had softened by 1817, when he accepted the classification of a single human species into three major races: the Caucasian (white), the Mongolian (yellow), and Ethiopian (black). He recognized extensive variation within these large groupings and thought that each had its own geographical origin, probably in the remote past and not long after the creation of the human species. Beyond this, though, Cuvier did not show a great deal of interest in human beings as a subject of study, although he reflected the beliefs of most of his colleagues in asserting that Europeans had larger heads, and thus more voluminous brains and higher intelligence, than other races. His major influence on students of human variety came more indirectly, as at various points in his prolific writings he argued energetically against the Great Chain of Being—a construct that was, of course, very evidently at odds with his succession of catastrophes.

With the exception of a few distinguished human anatomists such as Samuel von Sömmerring (who famously dissected and compared black and white cadavers) and Pieter Camper (remembered as the quantifier of facial projection in human skulls, but also an egalitarian when it came to racial differences), scientists writing on race up until the last quarter of the eighteenth century had mostly been general natural historians, seeking principally to situate human variety within the wider picture of nature. The first influential scientist to more or less make a career out of the study of human diversity (though he was also an expert on the platypus) was Johann Friedrich Blumenbach, a German physician who spent virtually his entire professional life at the University of Göttingen, rarely venturing out into the world beyond whose denizens so fascinated him.

Blumenbach was an avid collector of human skulls from around the globe and founded the science of cranial measurement that became such a preoccupation of human biologists in the nineteenth century. First published in 1776, his thesis *On the Natural Varieties of Humankind* rapidly became the definitive work on the subject, later influencing such luminaries as Lamarck. Blumenbach shared Buffon's view that the human species was truly unitary, while also agreeing with him that species could change internally. As translated into English, his resulting conclusion that "the causes of degeneration are sufficient to explain the corporeal diversity of mankind" may have involved an unfortunate terminology (the German term translated as "degeneration" might equally have been translated as "divergence"), but it did stem from the twin beliefs that species could diversify and that the various human races had all sprung from a single source. By the term "degeneration" Blumenbach more or less meant "the development of variety within species," and as used by him it carried no moral or judgmental overtones—although it did suggest a lingering essentialist feeling that the original type was the ideal and that departures from it somehow represented a form of deterioration, even when they involved specific accommodations to ways of life and environmental circumstances.

In the first edition of *Natural Varieties* Blumenbach recognized the four geographical groupings of human beings that Linnaeus had listed. But by the time the third edition was published, in 1795, he had concluded that there were, in fact, five human varieties. Each of these was identified with a particular geographic region. First among them was the white, rosy-cheeked Caucasian, so named because Blumenbach believed that the ancestors of Europeans had emerged in the Caucasus, between the Black and Caspian Seas. The Mongolian variety, yellow and straight-haired, embraced the inhabitants of Asia except for its Arctic peoples and those of India and the Malay Archipelago. All Africans except those in the far north belonged to the black, curly-haired Ethiopian variety; the copper-colored and straight-haired American variety included all the native inhabitants of the New World except for the Eskimos; and finally, the tawny-colored and luxuriantly haired Malay variety embraced today's Indonesian and Pacific peoples.

Yet while he was prepared to give names to these "principal" varieties of humankind, Blumenbach also perceived numerous "insensible transitions" among them. Physical boundaries between adjacent populations were not sharp, and to Blumenbach this was further evidence that all known humans belonged to a single species. Within that species he perceived two trajectories of change. As he saw it, the process of "degeneration" that had taken place among human populations in response to environmental stimuli had proceeded from a common "Caucasian" type, through the Malayan toward the Ethiopian on the one hand, and through the American toward the Mongolian on the other. And while in his earlier days he had hinted that the physical characteristics of the various human varieties might indeed be correlated with temperamental or aptitude differences, by the time of his third edition he was fully convinced of the equal excellence of the races.

Up until the early nineteenth century, then, scientists, philosophers, and even theologians had mostly performed quite impressively in broaching the mysterious issue of human variety, even in the absence of the explicitly evolutionary framework that would eventually provide the key to it. Empiricism reigned, and commentary upon regional differences was quite restrained, although a lot of ink was spilled in arguing about the relationship, or its lack, between climatic features and physical variables. Most significantly, natural historians especially showed little inclination to rank the human races in any particular order, even as few showed any reluctance to characterize them in terms of temperament, as well as of physical features. What's more, virtually everyone at the turn of the nineteenth century was a monogenist, believing that all living human beings had a common origin, belonged to the same species, and partook of the same essence.

POLITICS INTRUDES

Within a few decades this changed quite dramatically as a result of several factors, most of them political rather than scientific. Among more strictly scientific influences was the information flooding into Europe about the incredible diversity of the living world. This created a new intellectual environment in which older classifications demanded reinterpretation. At the same time biblical authoritarianism was weakening in some countries, including France and the United States. Then there was the long-lasting craze for the "science" of phrenology, invented by the German physician Joseph Gall in the last decade of the eighteenth century. Phrenology involved the belief that character traits were governed by specific areas of the brain, which imprinted themselves on the corresponding areas of the bony skull. The external form of the skull was thus an indication of the personality that dwelled within, and the resulting typology was applied to crania from all over the world, inviting the psychological stereotyping of exotic populations even as the more exacting science of cranial measurement was becoming popular in parallel.

A far more important influence, though, was the political scene. As the nineteenth century progressed there was a huge increase in imperial activity, as the greater part of the Old World came to be colonized by European powers. Driven by commercial and political competition among a small group of European nations, this process of colonization—and the associated practices of slavery and slave trading, which drew in the New World—nonetheless also demanded a more abstract justification. This was often sought in claims for the inherently inferior or "less developed" character of subjugated peoples. High mortality rates among Europeans in far-flung places as a result of colonial activity (few seemed to care much about the effects on local people of diseases brought by the newcomers) also helped break down ideas of geographical adaptation. For the fact that Europeans were dying rather than changing under such circumstances was often taken as evidence that climate, the traditional explanation of human variety, was not by itself a sufficient cause of racial differences. And if this were so, the differences between the races were more profound than could be accounted for by the superficial processes of adaptation, opening the door for a resurgent generation of polygenists to argue that the races were in fact different species, or were at least so profoundly differentiated as to indicate separate origins.

One of the leaders of this last school of thought was Samuel Morton, a Philadelphia physician and anatomist who around 1820 began amassing and studying a huge collection of human skulls from around the world. This became the basis for his treatise *Crania Americana*, published some twenty years later. A devout Quaker, Morton was disturbed by the fact that, while the authoritative Archbishop Ussher had calculated that the world was created in 4004 BC, the depictions of apparently fully formed black and white human races in Egyptian art were only a thousand years younger. A thousand years was an impossibly short period of time for the differentiation of the races we know today, especially given Morton's conclusion from his skull studies that they had not changed at all in the following five thousand years. There was simply no time in this scheme for humans to have become differentiated, by Blumenbach's "degeneration" or by any other known mechanism. In light of this, Morton found it "reasonable" to conclude that the Creator had "at once" created each individual geographical variety of mankind, specifically for the environmental and "moral" circumstances in which it was going to live.

Morton specifically avoided going into the matter of whether the varieties of humankind represented separate species, preferring throughout to allude to "races"—although he never quite got to grips with exactly what races were in principle, "primal varieties" being about as close as he came to a definition. This omission was almost certainly due to the difficulty Morton experienced in reconciling his polygenist conclusions with his belief in the biblical account of Creation, but elsewhere the longstanding confusion between "races" and "varieties" was beginning to be sorted out. Increasingly, "varieties" were becoming viewed as ephemeral, reversible divisions within species, while "races" were becoming primary divisions, with long histories and staying power. In this way Kant's notion of the permanence of race began to reassert itself. And since Morton's successors energetically promoted the Kantian position, it is basically to Morton that we can trace back the term "race" as used in the English-speaking world over the past couple of centuries.

Through an elaborate and painstaking procedural study of the crania in his collection—using techniques, largely relating to brain size, whose propriety has both been attacked and defended by recent writers—Morton effectively substantiated the five-fold scheme ultimately adopted by Blumenbach. He went well beyond anything Blumenbach contemplated, however, not only by subdividing his five races into twenty-two "families" but also by ranking them, as in the Great Chain of Being. At the bottom lay the Africans, among whom were to be found the "lowest grade of humanity"; at the top were Caucasians such as Morton himself, with the "highest intellectual endowments." The rest fell in between, displaying various character traits both good and bad. The deficiencies of the analyses that led to this ranking have been widely publicized lately, and we don't need to reiterate them here. However, it does seem that Morton's conclusions were not arrived at in the service of any particular political goal; they resulted from unconscious rather than from intentional bias.

But there they were, and after his death in 1851 Morton's ideas were seized on by apologists for slavery. Morton himself had been wary of polygenist ideas, because he had difficulty in reconciling them with the evident monogenism of Genesis. His professed disciples George Gliddon and Josiah Nott had no such scruples. In 1854 they published *Types of Mankind*, in which they made a specific argument for the separate creation of the races and for the inferiority of blacks to whites. This provided a tailor-made argument in defense of the southern proponents of slavery, a practice that was otherwise morally indefensible in a country of which the founding document asserted that "all men are created equal." Not that this was necessary in any practical way; an entrenched economic system, the heavy weight of social tradition, and the biblical curse of Ham were by themselves more than enough to sustain an arrangement in which the abuse of human freedoms was essential to the maintenance of economic prosperity. And while many slaveholders were doubtless happy to have such additional justification as Nott and Gliddon's for the unpleasant economic system that supported them, southern slave-owning society as a whole seems never to have felt it necessary actively to seek scientific justification for the practices that sustained it.

At an academic level far more exalted than that of the polemicists Gliddon and Nott, the polygenic notion was supported by Louis Agassiz, a distinguished professor at Harvard. A Swiss "spiritual disciple" of Cuvier, Agassiz was famous for his recognition of the geological evidence for a series of "Ice Ages" in Europe, and he had come to America in 1847 believing in Cuvierian multiple creations of fossil faunas. He also accepted, however, that the final creation had been followed by a unique single origin for humankind, an event that in itself proved the fundamental distinctiveness of humans from the rest of nature. Once in the multiethnic United States his ideas changed, partly because of a visceral reaction to the black domestics he saw all around him and possibly not least because of the strong friendship he had forged with Morton. Agassiz rapidly came to believe not only that the various races had indeed been separately created but also that their inborn abilities were different. In an essay published in 1850, while advocating the abolition of slavery (largely because he felt that it encouraged unnatural miscegenation between two separate species which should not be mixing at all), he recommended that relations between whites and blacks should be guided by "a full consciousness of the real difference existing between us and them, and a desire to foster those dispositions that are marked in them, rather than by treating them on terms of equality."

Just as Morton's work influenced disciples as varied as the cerebral Agassiz and the crude populists Gliddon and Nott, Morton himself acknowledged a debt to the polygenism that had been emerging in France as the nineteenth century progressed. As the century began, most French naturalists broadly agreed with the physiologist Georges Cabanis that "if human races did not continually interbreed . . . the physical conditions of each would be perpetuated from generation to generation," a position that uneasily combined an essentialist view of races with a monogenist conclusion. But during the 1820s the naturalist Julien-Joseph Virey tipped the debate back toward the polygenist side, with the suggestion that "Blacks" and "Whites" belonged to separate species. This proposition was soon thereafter elaborated by the physician Louis-Antoine Desmoulins, who ultimately came to believe that each race had an innate set of properties and, like species, individually pursued its own inner destiny. Underlying all of this were the writings of William Frederic Edwards, who had argued from the early years of the century in defense of the fixity of morphological types and their associated "moral characters." Most significantly of all, however, Edwards ultimately shifted the debate from the level of the individual to that of whole populations, and even nations. As he wrote in 1845, "we cannot really understand the moral character of a nation, if we do not understand that of the races which constitute it," thus generalizing the discussion from biology to sociology and history.

In purely political terms, this was probably where the argument belonged, since racial attitudes in society had never been free of those behavioral stereotypes that had already been reflected in the earliest biological classification by Linnaeus—and had, indeed, been inherited from similar attitudes among the Ancients. Of course, racism is a close relative of its equally ugly cousin xenophobia, born as much out of fear of strangers (especially strangers living in one's midst) as by guilt over exploiting them. As far back as 1714 Francis Atterbury, Bishop of Rochester, had complained that "The old, honest English stock" was being diluted by asylum-seeking members of "a different and base species." Who were these appalling aliens? German refugees from the Rhineland, who might easily have passed for native English (whatever that might have meant), at least until they opened their mouths. This is just one example of many, though Atterbury's grumblings were much milder than Ferdinand and Isabella's expulsion of the Jews from Spain had been a little over a century before, or than any number of wholesale massacres over the many centuries of recorded history.

Particularly in the first half of the nineteenth century, European and American intellectuals fervently discussed the significance of interbreeding among different geographical varieties of humans, and a huge and often conflicting literature accumulated on the subject. Early on, much of the debate was conducted within a relatively detached scientific context, but an equal amount of energy soon became devoted to political ends. The confusion that resulted from a welter of inconsistent reports on hybridization among various groups of humans was reflected in varying social attitudes in different places. Even where institutionalized slavery reigned (it may be significant that Edwards had been born in Jamaica, to a plantation owner), attitudes revealed considerable ambivalence. In the south of the puritanical United States a rigid social classification rapidly implanted itself, whereby anyone with any demonstrable African ancestry whatsoever became legally defined as black (the "one drop of blood" principle, a notion actually dating back to colonial times). In contrast, in Brazil, where nineteenth-century cruelty to slaves was if anything even more horrible than in the American South, a more relaxed attitude prevailed toward racial and cultural classification.

Apart from rapidly becoming an entrenched way of life, slavery was of course the center of an entire economic system, one that allowed short-term prosperity to override moral, scientific, or any other considerations. To its proponents, slavery needed no justification beyond their own affluence, and such fateful legal decisions as the one reached in the American Dred Scott case of 1857—which affirmed that blacks were not citizens but property and "had no rights that the white man was bound to respect"—provided their own bizarre rationale for the system. But such legal notions were also in tune with deeply ingrained social prejudices that valued "us" over "them" in an entire hierarchy of contexts; and, to that extent, they benefited from even the mildest support among scientists and other savants for the idea of a polygenic origin of humankind.

French science had provided Samuel Morton with early inspiration, and he repaid the debt in kind. Perhaps the most influential voice in French anthropology in the mid-nineteenth century was that of the physician and neuroanatomist Paul Broca, a founder of the Anthropological Society of Paris. Broca accepted Morton's basic assumptions and enormously developed his craniometric methods. In 1855 Broca clearly showed his hand by denying that "the polygenist doctrine assigns to the inferior races of humanity a more honorable place than the opposite doctrine" because "to be inferior to any other man either in intelligence, vigor, or beauty is not a humiliating condition." Yet there was also nuance: in 1862 Broca echoed Morton's conclusion that the human races were "primordial," but at the same time cautioned against the recognition of "racial types." The same point was neatly made in 1876 (with a subtle overtone) by Broca's student Paul Topinard: "A people is what is seen before the eyes or what history reveals; a race is what is looked for and is often assumed." Here was one of the first explicit intimations that race might be an intellectual rather than a biological construct. In practice, though, Topinard and other members of Broca's school of anthropology were not in the least hesitant to make peremptory and often demeaning judgments about the racial types they perceived.

Indeed, in the mid-1850s the French aristocrat Joseph Arthur de Gobineau had already published his widely read *Essay on the Inequality of Human Races*. There Gobineau claimed that culture was inseparable from each of the three races he recognized ("black," "white," and "yellow"); that the white race, descended from pure Indo-Iranian

"Aryans," was superior to the others; and that race mixture led only to moral and physical deterioration. Gobineau saw this process of decline by admixture as already highly advanced in Europe and as the cause of many social woes. Gobineau's ideas were seized upon almost a century later by the Nazis—although, since Gobineau himself was not anti-Semitic, Hitler and his colleagues found themselves obliged to quote very selectively from his writings. Significantly, Gobineau made no attempt to represent himself as a scientist; he was neither a natural historian nor an anthropologist, and the fact that he wrote with such assumed authority on this subject shows just how blurred the boundaries of the study of human variation were becoming, as social, historical, and political polemic intruded into what had formerly been the province of natural historians and other scientists.

In Britain the story was rather different. Despite the early speculations of Lord Kames, polygenist ideas had never made much headway there, and early- and middle-nineteenth century authorities such as the anatomist William Lawrence, and later the physician and ethnologist James Cowles Prichard, stoutly defended the unity of mankind. Not that monogenist or polygenist ideas necessarily implied different racial attitudes in practice: during the nineteenth century European superiority was widely assumed, if less often articulated. But Britain in the first half of the nineteenth century was generally too conformist a society to harbor many who were ready to desert the Bible and promote polygenist ideas, although the works of polygenists from across the Channel were quite widely read in translation. A more potent intellectual force among the small group of liberal British thinkers of the first half of the nineteenth century was, in fact, Lamarck, a thinker whose ideas were greatly in tune with the ethos of unstoppable progress embedded in the relentless change of the Industrial Revolution. And they certainly affected the world view of Robert Grant, an anatomist at the University of Edinburgh who taught and deeply influenced the young Charles Darwin during his brief foray into medical studies in the late 1820s.

ENTER EVOLUTION

As the nineteenth century began, the idea was already in the air that organisms were capable at least of limited change. To Lamarck such change was limitless; to Buffon, it was limited to whatever was possible within species. And in the last decade of the eighteenth century Erasmus Darwin, Charles's grandfather, had even proposed that "the final course of [the] contest among males seems to be, that the strongest and most active animal should propagate the species which should thus be improved." But the framework within which we understand the differentiation of human races today was not yet in place, and biblical notions still held wide sway. All that was to change in short order, after the radical theory of evolution by natural selection—independently arrived at by Darwin and his younger contemporary Alfred Russel Wallace—was presented to a meeting of the Linnean Society of London in 1858. And once Darwin had published his exhaustive opus *On the Origin of Species by Natural Selection, or the Preservation of Favoured Races in the Struggle for Life* the following year, science finally had a clearly articulated view of the mechanisms governing change over time in natural diversity. What's more, it was entirely evident that humankind was just one more component of that diversity.

After the predictable initial outcry, virulent enough to thoroughly traumatize the retiring and highly sensitive Darwin, it is quite remarkable how rapidly his central idea of "descent with modification" was accepted by English society as a whole, and far beyond. Finally the "nested sets" of organisms that Linnaeus had perceived were placed in a rational perspective: they resulted from a process whereby an ancestral species could give rise to several descendant species, which could in turn give rise to others, in a forking series. Every living thing on Earth, humans included, was linked by genealogy to a single common ancestor that lived very long ago indeed. The descendants of that ancestor had diverged over vast amounts of time, by a process that Darwin called "natural selection," the constant action of which over the eons had given rise to the vast variety of organisms in the world.

Natural selection had an immediate appeal, analogous as it was to the familiar "artificial selection" that so rapidly formed new breeds of domestic animals. Indeed, once pointed out it seemed so intuitively obvious that, on hearing of it, Darwin's colleague and defender Thomas Henry Huxley famously berated himself with the exclamation, "How stupid not to have thought of that!" Within any generation of a reproducing population, individuals vary in numerous inherited features. Those that reproduce most successfully are the "fittest," namely those whose inherited characteristics best "adapt" them to the surrounding circumstances. With the passing of the generations the appearance of the population changes, as favorable variants accumulate and those less fit diminish and are ultimately eliminated. Eventually, members of the same lineage will differ appreciably from their forebears, just as related species will diverge indefinitely from each other. The vast amounts of time necessary for such changes to accumulate were no longer a problem by the time Darwin wrote, since by the middle of the nineteenth century geologists had long since abandoned the Ussher chronology and had concluded that the Earth was of effectively limitless age.

Along with Darwin's meticulous documentation, it was the articulation of a plausible mechanism for change that helped convince the mass of educated people that evolution had indeed occurred. Darwin himself avoided applying his theory specifically to human beings, doing no more than hint darkly in the *Origin* that "light will be shed on the origin of Man and his history." But everyone got the point, and after enormous initial outrage it was quite rapidly accepted by many that human beings indeed shared a common ancestry with apes, monkeys, lemurs, primitive mammals, and all the other denizens of the world. Ironically, in the longer term it was natural selection, rather than common origin, that found itself under attack from Darwin's fellow scientists. We will look more closely at the processes of evolution as they are now understood in the next chapter, but in the human context it was almost inevitable that Darwin's notion that humans and apes shared a common ancestor, commonly caricatured as "mankind's descent from the apes," should be seized upon by apologists for the slave trade in a new version of the polygenist account of human origins.

To put things in historical perspective, *On the Origin of Species* was published just a week and a day after the hanging of John Brown for fomenting the slave rebellion and his raid on Harper's Ferry. Conflict over the slavery issue was dissolving the United States into chaos, and the monogenist-polygenist debate was losing any semblance of scientific detachment, even to the east of the Atlantic. In his *Lectures on Man,* delivered in Geneva and published in London in 1864, the zoologist Karl Vogt, a former student of Agassiz (and, ironically, an opponent of slavery), vociferously joined the German anatomist Hermann Klaatsch's proposal that the races of man were descended from different species of ape, a cry that was taken up by numerous others with a more specifically political agenda. In England these included the physician James Hunt, who translated Vogt's lectures and whose own *The Negro's Place in Nature* appeared in 1864, its title a tasteless play on that of the monogenist collection of lectures, *Man's Place in Nature*, that had been published in the previous year by Huxley. A more specific transatlantic connection was drawn in 1868 by the pro-slavery and anti-suffragist James McGrigor Allan in his book *Europeans, and Their Descendants in North America*.

On the more strictly scientific front, and also in 1864, Alfred Russel Wallace tackled the "one species or several?" question head on. With Morton and his followers, he admitted "the permanence of existing races as far back as we can trace them." But he also concluded that this was only "negative evidence" for true long-term permanence and that "a condition of immobility for four or five thousand years, does not preclude an advance at an earlier epoch." Indeed, he reasoned that the high sociality and cooperativeness of contemporary humans had effectively negated natural selection upon humankind's physical features, including those by which we recognize race. Natural selection in the remote past had made different human populations what they are physically; but, with the advent of complex society, "man's physical character had become fixed and immutable," as selection began to target our "mental and moral qualities" instead. Wallace thus concluded that the anatomical differences among the races must have been very ancient in origin, dating from a time when an ancient "single homogeneous race . . . gregarious but scarcely social" was still subject to natural selection on its physical features. This in no way affected mankind's essential unity, which was in fact greatly reinforced by its perceived common escape from the "great laws which irresistibly modify all other organic beings." But it did admit a vision of future struggle among the races, in which "lower" forms, those less strongly selected for mental quality, would inevitably succumb.

Eventually, goaded by the abuse of his theory by the likes of Hunt and Allan, and probably less than entirely happy with Wallace's views, the exceedingly reserved Darwin himself entered the fray with his two-volume opus *The Descent of Man*, published in 1871. By heritage, experience, and inclination Darwin, grandson of the pottery magnate Josiah Wedgwood, creator of the famous "Am I Not a Man and a Brother?" cameo that had become the unofficial emblem of the abolitionist movement in England, was deeply opposed to slavery. But, still in shock from the largely hostile initial reaction to the *Origin of Species,* in which he had so assiduously avoided referring directly to humans, he was initially highly reluctant to take on the explosive subject of humankind. Indeed, he found innumerable reasons for delaying the project. Less the treatise on human evolution suggested by its title than a plea for the unity of mankind on the one hand, and a showcase for Darwin's theory of sexual selection on the other, *The Descent of Man* was thus eventually published at a time when the U.S. Civil War had already been settled in favor of the anti-slavery forces.

In many ways the founding document of modern biological anthropology, *The Descent* reads as a curious amalgam of down-to-earth monogenism—"all the races agree in so many unimportant details of structure and in so many mental peculiarities, that these can be accounted for only through inheritance from a common progenitor; and a progenitor thus characterized would probably have to rank as a man"—and robust Victorian stereotyping of non-English peoples, including of course the "careless, squalid, unaspiring" Irish.

In his relationships with individuals of all backgrounds Darwin was the most considerate and caring of men; that he was happy to trade in characterizations of this kind is evidence that, while by the fourth quarter of the nineteenth century the battle against biblical polygenism had effectively been won in England by the evolutionists, Victorian science as well as society was still rife with national and racial stereotyping. Certainly, polygenist views were to rear their heads again in England, even among professed supporters of evolution. Perhaps the most interesting aspect of *The Descent*, though, is that in its many pages Darwin failed to find his mechanism of natural selection an adequate explanation of the varieties of humankind. Too many differences were simply unattributable to the climatic and other environmental influences that had traditionally been drawn upon to explain them—hence, of course, the attention he paid to sexual selection (basically, mate choice by females) as an agent of change.

ACROSS THE CHANNEL

Things were very different in Germany, where notions of change were also in the air in the first half of the nineteenth century, in the guise of the movement known as *Naturphilosophie*. This Romantic vision of a transcendent reality saw humans as essentially one with nature, which was animated by a supernatural force. The resulting dynamic expressed itself in the unfolding of a sort of divine plan for the world, in which every species had its place. Georg Bronn, the distinguished zoologist who translated Darwin's *Origin* into German, subscribed to this outlook and was thus not entirely in sympathy with Darwin's convergent yet much more materialistic views, particularly as they were generalized by others to human beings. Not so his younger colleague Ernst Haeckel, also a devotee of *Naturphilosophie*, who was powerfully affected by Darwin's work and who used the view of change it espoused as the basis for his own books *Natürlische Schöpfungsgeschichte (History of Creation)* and the wildly popular *Die Welträthsel*, translated into English in 1899 as *The Riddle of the Universe*.

Haeckel was certainly the most energetic continental European contemporary advocate for Darwinism in its broadest sense, and particularly for the mechanism of natural selection, yet in sharp contrast to Darwin he was no monogenist, seeing the differences between the modern races as equivalent to those between other mammal species. He was troubled in a minor way by the lack of well-defined boundaries among the races, and by the obvious interfertility among all humans, but although he believed that all humans were descended from an ape-like ancestor, Haeckel was also convinced that the various modern human "species" had originated from different descendants of this ancestor in distinct parts of the world. What's more, he believed that language, the "special and principal" attribute of humanity, had been acquired independently in each of the resulting lineages.

Beyond this, Haeckel believed passionately in the inequality of the races, building on Gobineau's notion of "Aryan" superiority. He identified the Aryans with the true German *Volk*, a tall, blond, blue-eyed race with a transcendental unity and deep roots in the landscape of the Fatherland. This pure race was destined to be victorious in the struggle against lesser varieties of humanity. The tall, blond, and blue-eyed Haeckel's propagation of the Aryan myth was adamantly opposed by his former teacher, the pathologist Rudolf Virchow. Virchow was the only German biologist of comparable standing to Haeckel, if shorter of stature and sallower of complexion, and a titanic battle between the two raged on this and other issues until Virchow's death in 1902. By means of an extensive survey of German schoolchildren, Virchow demonstrated that only a minority fit Haeckel's notion of the exemplary German, but, sadly, this finding did little to weaken the Aryan ideal in the public mind.

Almost equally unfortunately, in the process of rejecting Haeckel's Aryanism Virchow also rejected Darwin's notion of evolution, thereby throwing out the baby with the bathwater. Indeed, it is probably out of blind anti-Darwinism that Virchow dismissed the Neanderthal skeleton, the first human fossil ever to be recognized as such, as nothing more than the bones of a diseased modern human. Both Haeckel and Virchow were first-rate scientific minds, and each made fundamental contributions to the progress of nineteenth-century empirical biology, Haeckel in the field of embryology and Virchow to an even greater extent in cellular pathology and the theory of disease. But when it came to race and evolution, each co-opted their science in the service of broader political, philosophical, and social agendas, and thereby distorted both. By the time they were through, science and politics had become inextricably intermixed, and for a long time there was no going back.

SOCIAL DARWINISM

The English have never really known quite who they are, so the response across the North Sea to Darwin's revolutionary new perspective was bound to be rather different from the ebulliently nationalist Haeckel's. Nonetheless, developments there were in some ways equally disturbing. Rather than generalize from the success of the individual to that of the group as a whole, the English emphasis remained on the individual, where Darwin had placed it. One result, somewhat to Darwin's own dismay (he thought it the result of facile, unsubstantiated theorizing), was the development of a school of sociological thought that was labeled "social Darwinism" by its leading practitioner, the philosopher and social theorist Herbert Spencer. Spencer it was who famously coined the term "the survival of the fittest," a snappy if rather inaccurate description of the evolutionary process that Darwin himself later adopted, at Wallace's urging, as a tag line readily grasped by the public.

Spencer had actually entertained a theory of cosmic and organic change even before Darwin published the *Origin*, and once the Darwinian cat was out of the bag, he set about applying the survival of the fittest theme to society in general. Haeckel, who had once declared that "politics should be biology writ large," was busily supporting policies that promoted the Aryan nature of the German population; and in stressing the importance of natural selection within societies, Spencer followed a rather similar line, albeit one shorn of nationalism. As far as the enormously influential Spencer and his followers on both sides of the Atlantic were concerned, social policies aiding the poor and the infirm only weakened the population by allowing the unfit to proliferate. Clearly, it would be much better for society if government were simply to stand aside and let nature take its course. This amounted to the advocacy of unfettered capitalism, and in the midst of an industrial revolution that entailed enormous social upheaval and a rewriting of the relationships among and within the upper and lower classes, it found fertile soil in which to flourish among the more privileged strata of society throughout the English-speaking world.

What's more, from Spencer's laissez-faire stance it was not much of a leap to eugenics, a movement given its name in 1883 by Darwin's mathematically inclined younger cousin Francis Galton. Once defined as "the self-direction of human evolution," eugenics sought to improve the species by promoting the breeding of more intelligent and otherwise admirable individuals, while its counterpart, negative eugenics or dysgenics, attempted to suppress breeding among the feeble of mind and body. Such goals were entirely in tune with the progressivist Victorian ethos of the day, and they were first articulated in a rosy idealist spirit. Appalled by the "nature red in tooth and claw" aspect of the natural selection that Spencer wanted to see let rip in the human population, late in life Galton summarized his views as follows: "Man is gifted with pity and other kindly feelings: he has the power of preventing many kinds of suffering. I conceive it to fall well within his province to replace Natural Selection by other processes that are more merciful and not less effective." Nonetheless, apart from death and taxes the law of unintended consequences is possibly the only ironclad universal of human existence, and the paternalistic Galton had unwittingly opened the door to some of the most horrible excesses in recent human history. This was the slipperiest of slopes, and while Galton's own interest was in fostering excellence in society and the human species in general, in the event—apart from the short-lived 1930s Nazi experiment that attempted to breed a pure Aryan master race—dysgenics made a much greater impact than eugenics did. Well before the end of the nineteenth century a drive for forced sterilization of those deemed undesirable was under way on both sides of the Atlantic, and it continued periodically to raise its head until it was ultimately extinguished only in reaction to the Nazi horrors preceding and during World War II.

Meanwhile, the perpetually curious Galton was intrigued as much by the differences among the human races as by how to achieve improvement within them. By the time he took up the topic, subjective ranking of the human races had become routine among those who wrote on the subject; and even the monogenist Huxley, for example, had been happy to allude to "lower" and "higher" races in *Man's Place in Nature*. But Galton had a new angle in seeking ways to quantify such ranking. Using the rather dubious criterion of how productive each race was of individuals he considered outstanding, he generated statistical analyses suggesting that, on average, the English led the pack among modern people, with Africans scoring well below them and Australian aborigines lower yet. Nothing to raise any contemporary British eyebrows there.

But highest-scoring of all in Galton's ranking were the ancient Athenian Greeks (who had also actually scored highest, coming in above modern Europeans, in Pieter Camper's facial-index studies two centuries earlier). And this, of course, suggested something rather unflattering about the state of the modern world. Nowhere was anxiety about such perceived deterioration of the stock more widely shared than in the "melting pot" of the United States, where peoples of hugely diverse geographical origins were mixing as never before and raising deep fears in the minds of some over the dilution of the founding "white race." A leading spokesman for worries of this kind was Madison Grant, author of *The Passing of the Great Race*, published in 1916. An energetic amalgam of racism, stereotyping, paranoia, and pseudoscience, *The Passing* sold like hotcakes. Delving for distinctions within distinctions, Grant divided the traditional "Caucasian" race into three: Nordics, Alpines, and Mediterraneans. And he identified the Nordics (a cultural and biological group centered in Scandinavia but with ultimately Teutonic origins) as the engine of all civilized society. The Nordic race had acquired its admirable features in response to the bracing and challenging environmental conditions in which it had evolved, and Grant deplored its adulteration by admixture with lesser races, including Alpines and Mediterraneans. His solution? The use of eugenic (or rather, dysgenic) methods to eliminate "undesirable" characteristics and "worthless race types" from the population.

DESCENT INTO HORROR

While Grant's "scientific racism" received enthusiastic support from eugenicists such as the zoologist Charles Davenport, and eugenics itself captured the public imagination and the legal system to the extent that, ultimately, some sixty-five thousand Americans were forcibly sterilized, his beliefs also found opposition from the beginning. One energetic adversary was Franz Boas. Through his many students, Boas and his ideas ultimately came to dominate U.S. cultural anthropology in the middle part of the twentieth century, but it was a very lonely row to hoe at first. To Boas, born in Germany to a secular Jewish family and influenced as a student by Virchow, the study of "race" in the traditional sense was irrelevant at best, and dangerous at worst. Culture, not biology, was the driving force of human nature and the determinant of differences among societies. It was a grievous error to link the two and to associate different races with specific temperaments or aptitudes, as was so carelessly and routinely done in the United States of Boas's time. Recognition of the fundamental importance of culture to humanity was already implicit in the work of earlier U.S. anthropologists such as Lewis Henry Morgan, but making it the cornerstone of anthropological belief involved doing away entirely with the ancient typologies, and required a Herculean effort. So great an effort, indeed, that even in the face of events unfolding in Germany, Boas initially found his task almost impossible.

Soon after Adolf Hitler became German chancellor in 1934, Boas became concerned about the potential fate of Jews in the Nazi state. Starting even before the establishment of an Office of Racial Policy in that year, a raft of new laws had begun the process of anti-Jewish discrimination in Germany. Gradually, Jews were expelled from the medical profession, in which they had excelled, until in 1938 all medical licenses issued to Jews were revoked. In 1935 the Nuremberg Laws demanded that all those intending to marry be racially and physically certified, complementing the 1933 Law for the Prevention of Genetically Diseased Offspring. By 1939 the carnage had begun, with the killing of institutionalized mentally and physically handicapped children. This was soon extended to adult psychiatric patients, and then to Jews, homosexuals, Gypsies, and anyone else considered undesirable, culminating in the horrors of the "Final Solution." The involvement of science—or rather, pseudoscience—with social and racial policy had reached its nadir.

In 1930s England a variety of colleagues both shared and voiced concerns similar to those of Boas. Among them were the cream of the British science establishment, including the great mathematical geneticist J. B. S. Haldane (curiously enough, the author of a book in which he had predicted an early form of genetic engineering), the evolutionary biologist Julian Huxley, the zoologist and statistician Lancelot Hogben, and the anthropologist Alfred Haddon. In the United States, however, few academics—or, for that matter, Congress, which Boas tirelessly but unsuccessfully urged to protest—seemed at the time to care much. Boas cast around widely among prominent U.S. biologists to find support, but among non-Jewish colleagues he found only the Harvard physical anthropologist Earnest Hooton willing to take a public stand alongside him. The pair made a bit of an odd couple, especially since the much younger Midwesterner Hooton had a deep interest in racial classification, was attracted by eugenics, and was more than a little conflicted about the relationship between culture and achievement among Jews. He also served (along with Davenport) on the notorious Committee on the Negro, a joint enterprise between the prestigious American Association of Physical Anthropologists and the National Research Council. Set up in 1926, the committee eventually reported eleven years later that "the negro race is phylogenetically [evolutionarily] a closer approach to primitive man than the white race." Balancing this, however, was Hooton's firm egalitarian belief that "there are no racial monopolies either of human virtues or of vices" and his increasing unease about the corruption of anthropological science that was continuing apace in Germany. In any event, the collaboration between Boas and Hooton had little practical effect. At its annual meeting in 1939, the American Association of Physical Anthropologists considered a motion to dismiss any link between human physical and psychological-cultural differences, to reject "Aryan" and "Semitic" racial categories, and to deplore racism. It was diverted to a committee and was never passed. And right up through the middle of World War II, at least thirty states maintained laws banning miscegenation, involving an incredible welter of different "racial" groups including Mongolians, Kanaks, Malays, and a host of others, as well as the traditional "blacks" and "whites."

Regrettably, at least in an intellectual sense the dreadful events in prewar Germany that were so conspicuously ignored in the United States can be traced back fairly directly to Haeckel's embrace of Aryanism and his particular interpretation of social Darwinism. Having said this, of course, we have to emphasize that the linking of Darwinian evolutionary thought with the Nazi atrocities—an allegation occasionally aired even today—is based on totally false premises. Before his death in 1882 Darwin generally expressed approval of Haeckel's promotion of his evolutionary ideas (though his wife was less than impressed by the man himself). But there is little doubt that Darwin would have been distressed by the vigor of Haeckel's extension of his ideas into the social and political realms in the last two decades of the nineteenth century and beyond. And there is certainly no logical or scientific connection whatever between the idea that the diversity seen in the modern world came about via descent with modification and the notion that artificial—or even natural—selection should be allowed to rampage in modern societies.

But while Darwin mistrusted the notion of an interface between science and politics, to the point of conspicuously avoiding public debate of any kind, to Haeckel there could be no separation between the two. Indeed, in 1904 the German founded a popular organization he called the Monist League, a rabidly anti-clerical society that for two decades widely and energetically preached that Aryan superiority was under threat in Germany from dilution by inferior races. After the end of World War I this message had an amplified appeal to a public deeply aggrieved by the economic hardships imposed on the country by the terms of the armistice: the halt and the lame and the racially inferior were publicized as a terrible burden carried by the idealized German worker, who despite heroic efforts could barely feed his family. Something should be done! Ironically, the league was ultimately closed down by Hitler because of its support for other social goals such as feminism and pacifism; but by that time the *Rassenhygiene* it urged had already become an official policy of the National Socialist Party. Even the Nazis' growing obsession with the Jews may have had its roots at least partially in Haeckel's own anti-Semitism, since it appears that Hitler lifted some of his own anti-Semitic rantings more or less directly from the pages of Haeckel. Never before had science been so deliberately and destructively co-opted in the service of a political world view.

POLYCENTRIC EVOLUTION

One of those forced out of Germany by Nazi policies was the very secular Jew Franz Weidenreich, an anatomist of distinction who left Heidelberg for the University of Chicago in 1935, and the following year transferred to the Peking Union Medical College in China. In the process he became honorary director of the Cenozoic Research Laboratory in Beijing (then Peking), the center of research on the then newly discovered Peking Man. In the course of monographing this extraordinary collection of early human fossils (now thought to be about half a million years old), Weidenreich developed a "polycentric" model of human evolution. Particularly in his later years, he was very clear about the unity of the human species: "Anthropologists all over the world have agreed that living mankind classified by Linnaeus as 'Homo sapiens' represents a species in the taxonomic sense and that its main regional variations have to be considered as races or subspecies." Those "main regional variations" approximated to four out of the five human species postulated by the polygenist and highly racist Canadian geneticist R. Ruggles Gates in 1944: Australian Aborigines, San "Bushmen," Africans, Asians, and Europeans (Weidenreich's scheme left out the San). Whatever one might think of Gates and his species, if one were to twist an anthropologist's arm today to name the "major races," these five groupings are those he or she would probably come up with. Weidenreich convincingly demolished Gates's arguments for seeing the regional variants of humankind as separate species, and he was particularly scathing about the hugely complex and hairsplitting profession that, following Morton and Broca, had sprung up around the measurement of human skulls: "Since Camper the anthropologists have expressed differences by measurements . . . but only a few took pains to test whether the differences were of any diagnostic value. Everyone . . . is aware of the lot of useless ballast which is carried through the literature and piles up further every day." Amen to that.

Weidenreich's own analysis suggested to him that all of the major human races were the ultimate offspring of a single "continuous evolutionary line" that ran from what we would now call *Homo erectus* (including Peking Man), through the

Neanderthals, to the people of today. But since he believed that he could trace various "racial" characters of the skull deep into the past in the areas of the world in which they still occurred, he concluded that each of those races had very deep evolutionary roots. In 1947 (the year before he died) Weidenreich, never a master of the most limpid English prose, wrote that there was an "already world-wide distribution of early [hominid] phases which transmuted into more advanced types by vertical differentiation, while they split into geographical groups by horizontal differentiation." Thus, all hominid fossils known at that time formed "a unity with a faculty to pass into other forms and to split at the same time into different racial groups. . . . [I]t is supposed that main racial groups of today developed in parallel lines from more primitive human forms."

This rather opaque statement was later taken as the foundation of the "single-species hypothesis," of which we will soon hear more, and its later refinement known as "multiregional evolution," which co-opts almost all fossil hominids into a single lineage that, over more than a million years, diversified by adaptation to local environments while remaining unified by interbreeding. Given the vast amount of time involved this seems, perhaps, a bit like trying to have one's cake and eat it, but it has been an influential point of view nonetheless. And though Weidenreich was an instinctive egalitarian who would have been appalled by the suggestion of any political overtones to his conclusions, he was nonetheless prepared to observe that some of his human lineages preserved more "simian stigmata" than others.

To put all this in context, it has to be borne in mind that Weidenreich's principally interwar career spanned a time when virtually all of the most influential students of the human fossil record were more greatly preoccupied by the differences among the races than by their similarities. Thus in the United States Earnest Hooton, for all of his later antiracist activities, had remarked in 1931 that "the differences between the several races are quite as much as usually serve to distinguish species in animals." Across the Atlantic Sir Arthur Keith, the most prominent British paleoanthropologist of the interwar years, shared Weidenreich's belief that "the facts available do not indicate that environmental conditions played a decisive role [in human evolution]," and he, too, invoked "orthogenesis" (the idea that evolution was "directed" by some inherent mechanism) to explain how modern humans had come to be. But he went considerably further. Beyond his 1936 suggestion that "throughout the Pleistocene period the separated branches of the human family appear to have been unfolding a programme of latent qualities inherited from a common ancestor of an earlier period," Keith went on to claim that "the future of each race lies latent in its genetic constitution." Indeed, following the events at the end of World War II, he became something of an apologist for the "patriots" who had been hanged at Nuremberg.

POSTWAR REACTION

The excesses of the Nazi period finally wrote an end to eugenics, at least in its established form. Following World War II scientists everywhere, even—or perhaps especially—in Germany, where many had found themselves under pressure to become "race scientists," recoiled from the unimaginable horrors inflicted in the name of racial classification and improvement. In order to find context, many reached back to a 1935 book, *We Europeans: A Survey of "Racial" Problems*, that Huxley and Haddon had published in England just as Boas was trying vainly to raise the alarm in the United States. In this sustained attack on "race science," both as practiced by Madison Grant in the United States and by the National Socialists in Germany, Huxley and Haddon argued that "pure races" were entirely hypothetical concepts, with no existence in reality. All attempts to classify human variety stumbled on exceptions and on questions of variation. Yes, of course "races" existed, in the broad sense that people from various parts of the world looked rather different; but when you examined this variation more closely, there were no clear demarcations, and any "racial" character you cared to name could be found in combination with virtually any other.

This theme was most energetically taken up by the London-born Ashley Montagu (*né* Israel Ehrenberg), who had begun studying anthropology with Boas at Columbia University in the 1930s. As early as 1942 Montagu published his most famous book, *Man's Most Dangerous Myth: The Fallacy of Race*. The title of this work perfectly reflected its content, and its popular success ultimately procured him a (very difficult) postwar job with UNESCO, as coordinator of an international panel charged with producing a statement on race to be approved by the United Nations. The historian Michelle Brattain has neatly characterized this project as "reflect[ing] postwar liberal optimism about the power of internationalism and science itself to prevent human tragedy." Yet for all the undoubted optimism, and for all the enormous disrepute into which Nazi excesses had plunged the whole idea of "scientific racism," consensus was not easy to find. Finally released in 1950, the UNESCO statement reiterated that all humans belonged to the single species *Homo sapiens*, and that this species contained

both genetically and culturally distinctive populations that were variable within
themselves. What's more, it clearly showed Montagu's (and ultimately Huxley's and
Haddon's) influence by suggesting that the politically discredited term "race" should
be discarded for such groups and replaced by the more neutral (if less accurate) term
"ethnic groups," stressing that "biological differences between ethnic groups should be
disregarded from the standpoint of social acceptance and social action. . . . [T]here is
no proof that the groups of mankind differ in their innate mental characteristics."

Despite its undisguised attempt to shift the notion underlying race from the biological to the sociocultural sphere, this was a pretty anodyne declaration. Nonetheless,
it raised a remarkable uproar among those who had not participated in its drafting,
and a new panel was eventually convened to produce a revised statement. Initially
Montagu was excluded from this group, but he eventually resumed his place on the
committee, as spokesman for the first version. The road to the second statement was a
rough one, as might have been expected of a reconstituted panel that embraced many
more points of view than Montagu's carefully chosen predecessor committee had. Indeed, due to internecine disagreement an initial draft was abruptly withdrawn on the
eve of its intended publication in late 1951. The final second statement was a rather
lumpy compromise, eventually allowing that when groups of humans showed "well-
developed and primarily heritable physical differences" from others, they could indeed
legitimately be called races. And on the matter of "innate capacity" versus "environmental opportunity" as determinants of IQ test results, the panel waffled.

The fuss over the UNESCO initiative, and the resulting uneasy settlement
among the various protagonists, reflected the tension that had developed during the
immediate postwar years between the old, essentially typological, way of looking at
human variation and the nascent "New Physical Anthropology." This latter movement,
exemplified to some extent by Montagu but more particularly by a former student of
Hooton's named Sherwood Washburn, rejected the traditional static, descriptive approach to human variation and evolution in favor of a more dynamic vision quite
closely allied to the one articulated in 1957 by the English physical anthropologist
J. S. Weiner, when he described *Homo sapiens* as "a widespread network of more-or-less
interrelated, ecologically adapted, and functional entities."

Washburn believed that biological anthropology should seek to integrate behavioral, functional, genetic, and population data to create a seamless picture of human
biocultural history. In short, he wanted to marry biology and anthropology in a way

in which they never had been joined before. And he had good reason for this desire. This was because paleoanthropologists (most of whom had backgrounds in the highly focused study of human anatomy) had been in the habit of issuing authoritative statements about human fossils while knowing little and caring even less about evolutionary theory, while human biologists had traditionally been more interested in categorizing types than in understanding the genetics that underwrote variation within and among the populations they scrutinized. Washburn wanted a bigger picture, and one reason he wanted it was that, while a junior faculty member at Columbia University, he had fallen under the influence of the geneticist Theodosius Dobzhansky. In turn, Dobzhansky was one of the primary architects of the intellectual edifice known as the Evolutionary Synthesis and author of perhaps its most influential single document, *Genetics and the Origin of Species,* initially published in 1937. In this work Dobzhansky was much more interested in the distribution of human variables such as blood group frequencies than in the populations into which they were packaged, and in general he championed the view that a race was no more than a "group of individuals which inhabit a certain territory and which is genetically different from other geographically limited groups."

THE SYNTHESIS

The half-century following the rediscovery of the basic laws of genetics in 1900 was a time of enormous ferment in evolutionary biology. Heredity had, of course, been at the center of Darwin's notion of natural selection, though Darwin himself hadn't known how it worked. But the basic functioning of heredity, via the discrete elements known as genes, was beginning to be understood, and geneticists, natural historians, paleontologists, and others were energetically seeking to discover how heredity and evolution intermeshed. Theories of evolutionary mechanism abounded (including the orthogenesis to which Weidenreich subscribed), but not until the late 1920s did consensus begin to emerge on how evolution actually worked, in the shape of the convergence, just mentioned, that became known as the Evolutionary Synthesis. We will look more closely at all this in the next chapter, and for the moment it suffices to say that the Synthesis was a rather reductionist formulation that emphasized the role of the genes in the evolutionary process. In its earliest expressions the Synthesis was actually quite nuanced, but by the time the postwar years rolled around, virtually all evolutionary transformation was regarded in the English-speaking world as a matter of changing gene frequencies in populations (lineages) of freely interbreeding organisms, under the guiding hand of natural selection. As selection tirelessly worked, such lineages altered insensibly in appearance, so that over enough time, and with the accumulation of enough differences, one species became another. Diversity in nature was similarly explained, as lineages diverged by selection after becoming divided by ecological and geographic changes. Evolutionary histories thus became synonymous with the dynamics of gene changes within populations, principally guided by natural selection, though the potential contribution of random factors ("genetic drift") was also admitted.

As we have noted, prewar paleoanthropology had operated in something of a theoretical vacuum; and the Synthesis had a profound effect on the course of the study of human evolution after Dobzhansky and Washburn jointly organized a 1950 conference at the Cold Spring Harbor Biological Laboratory on New York's Long Island (an independent research center that is today a leading center in molecular genetics but had, ironically, once hosted Charles Davenport's Eugenics Record Office). At this conference, various giants of the Synthesis undertook to point out to paleoanthropologists the error of their ways. Most significantly, the ornithologist Ernst Mayr, who had probably never held a hominid fossil in his hands, roundly declared not only that the entire human fossil record (which by then included several ancient tiny-brained forms, as well as Peking Man and its like, and the big-brained Neanderthals) could not only be fitted into a single evolving lineage, but that all of the fifteen-odd hominid genera then recognized as a result of incompetent systematics could in fact be squeezed into the single genus *Homo*. The message was clear: if the dazzling diversity of hominids known over a huge span of time amounted to no more than variation within a single genus, then the differences among modern human populations were beyond trivial. And beneath this message was a subtext, detectable in many contributions to the conference. As recent political history had shown, there was considerable practical danger in scrutinizing those human variations in excessive detail. For, whatever one said, it could be twisted and misconstrued in the service of politics to which it bore no relation.

Students of human evolution in the audience and elsewhere were clearly traumatized by Mayr's attack. Whereas it had previously been more or less routine to mindlessly baptize each new hominid fossil found with a new genus and/or species name, paleoanthropologists for the following couple of decades could barely bring themselves to utter a taxonomic nomen: diagrams of hominid evolution during the 1950s and 1960s were commonly decorated with the identifiers of individual hominid fossil specimens ("Steinheim," "Swanscombe," "Broken Hill") rather than with species names.

What's more, many paleoanthropologists took to heart Dobzhansky's mantra, echoed by Mayr, that the human "ecological niche" had been so "broadened" by the adoption of "culture" that it was inconceivable, even in principle, that more than one hominid species could ever have existed on Earth at any one time. In contrast, many of the paleoanthropologists' colleagues who studied modern human variation rather than fossils pushed on, undeterred and even inspired: if human populations vary, then there must be some reason for that variation. And the Synthesis pointed its finger directly at natural selection. What were the agents of that selection?

This preoccupation turned attention away from the whole individual or group to the individual anatomical features upon which selection was believed to act—skin color, body proportions, and so forth. But then again, if you were going to talk about selection, you were going to have to talk about populations. This is because, even if the individual is the target of selection, it is obviously the population that evolves in the sense of becoming somehow changed. What's more, if you were going to talk about populations you had to be able to define them; and if you could define them, you could give them names. Even Dobzhansky was by then prepared to pursue this logic. In his 1962 book *Mankind Evolving,* he made no bones about the fact that human races exist: "Race *differences* are facts of nature." The human species is "polytypic"; it consists, like most successful and widespread mammal species, of several physically and genetically distinguishable local populations that you could call either "subspecies" or "races." Biologically speaking, Dobzhansky said, there was no difference. Some of the between-population differences were probably due to natural selection, others to genetic drift. Critically, though, since the races were fully interfertile, "pure" races did not exist. Nothing here for anyone to take exception to, but it did bring things to something of an impasse, because Dobzhansky never quite finished his thought, something we will do later. Still, Dobzhansky made his key point crystal clear: "nobody can discover the cultural capacities of human individuals, populations or races until they have been given . . . equality of opportunity to demonstrate those capacities." This statement was both incontrovertible and strictly scientific, but it was also one that carried a strong social message.

POST-SYNTHESIS

Some biological anthropologists, such as the judicious traditionalist Stanley M. Garn of Antioch College, responded to the Synthesis and its anthropological expressions by suggesting that "race" was as good a word as any for human variants and that it was helpful to distinguish among "geographic-," "local-," and "micro-races" in searching for the adaptive meaning of the physical variants involved. Still, before long it came to be fairly widely perceived that scientists were uselessly tying themselves into knots by making increasingly fine divisions among human populations as coarser ones dissolved under scrutiny. Even such new wheezes as using blood-group frequencies in place of traditional morphological characters had only served to complicate things further. By its very nature, human variation across the globe defied classifications that recognized discrete boundaries at any level. And it thus defied any classification at all. Representative of a powerful trend toward the recognition of this reality in the post-synthesis United States was the genetic anthropologist Frank B. Livingstone, of the University of Michigan. Livingstone's hugely influential 1962 article *On the Non-existence of Human Races* urged not only an abandonment of the race concept but its replacement in physical anthropology by the study of the ways in which physical traits and their underlying genes were distributed in different environments around the world. As energetically advocated by Livingstone, the study of human variation thus became focused on the genetics of populations, rather than on the ways in which the physical characteristics of those populations were packaged, and on the analysis of continuous geographic gradients ("clines") in gene frequencies, rather than on boundaries among populations. Indeed, Livingstone went as far as to declare that "there are no races, only clines."

But not everybody agreed, least of all the University of Pennsylvania anthropologist Carleton Coon. Also a student of Hooton, Coon was one of the last anthropological generalists, who pursued a rather swashbuckling early career that combined elements of ethnography, archaeology, and physical anthropology. By 1962 he had become president of the American Association of Physical Anthropologists, a position from which he ostentatiously resigned when the association voted to condemn his distant relative Carleton Putnam's pro-segregationist book, *Race and Reason*. His reasons for resigning were complex, and by some accounts had much more to do with his disapproval of politics within the association (and indeed, of the association's embrace of politics) than with his support for Putnam's book. Still, Coon's connection to the blatantly racist Putnam did nothing to ease his relations with opponents when he published his volume *The Origin of Races*, also in 1962. Neither did its contents.

In this thick tome Coon examined the entire human fossil record then known for evidence bearing on the origins of the human races (he recognized Gates's five).

And he concluded that Weidenreich had been right: the origins of the five major geographical races lay deep in time. Still, Coon was keenly aware of the new perspective offered by the Synthesis, and he tried to blend the Dobzhansky-Mayr outlook with that of Weidenreich. He thus envisioned a scenario in which adaptation to local conditions had been offset by "genetic contact [among] sister populations." As a result, the unity of the human lineage was preserved even as, over long periods of time, adaptive characteristics became more pronounced in each geographic population.

Since Coon's model emphasized continuity, he needed a criterion for recognizing members of *Homo sapiens* in contrast to their *Homo erectus* ancestors. He chose to use brain size and placed the threshold at around 1,250 cc (the modern average is about 1,350 cc; that of those fossils he recognized as *Homo erectus* was around 1,000 cc). Coon then discovered that larger-brained human fossils appeared at different times in different places, leading him to conclude that the various geographic lineages of humans had crossed the line to *Homo sapiens* at different times. On the available fossil evidence Asians had done so first, although Europeans possibly squeaked by on the indirect basis of archaeological evidence (essentially unavailable at the time for the other groups). Based on the fossils that Coon had to hand, Africans came last—although with today's better record and larger perspective we know that *Homo sapiens* evolved first in Africa.

Predictably enough, Dobzhansky, Montagu, Washburn, and many others vilified Coon as a racist. Whether or not he was a conscious or an unconscious racist was later disputed among those who knew him, though the historian John P. Jackson has recently argued cogently that, at the very least, Coon "actively aided the segregationist cause in violation of his own standards for scientific objectivity." Coon was also excoriated as a poor biologist (how could one species independently evolve five times into the same descendant species, and what did a threshold based on one notoriously variable character mean anyway?). The debate rapidly became highly polarized, politicized, and personal, and the rancor it generated ultimately helped drive human biologists away from the study of race and toward the less politically charged study of variation in individual human traits, in the context of adaptation.

Emotion was hardly dissipated by Coon's publication in 1965 of a companion volume, *The Living Races of Man,* in which he did his best to appear measured while succeeding only in adding fuel to the fire. Indeed, even in the unlikely event that Coon had had the most admirable of intentions, things could hardly have been otherwise;

this was after all the year of the march on Selma and the violent events of Bloody Sunday. But while most of his colleagues railed against his books, some rallied to his support, although not necessarily on strictly scientific grounds. As quoted by Coon in his autobiography, the eminent physical anthropologist Alice M. Brues praised his bravery in going against the grain: "in this era of the New Prudery, r-ce . . . has replaced s-x as the great dirty word. . . . [J]ust as our Victorian forebears did not entirely succeed in sweeping sex under the rug, so we cannot conceal forever the fact that people in different parts of the world do not look entirely alike. Better to learn it in school than from vulgar companions around the corner!"

Despite all the high drama and extreme politicization of the notion of race, some scientists—who conspicuously disassociated themselves from Coon, harking back instead to the innocent purity of the racially persecuted Weidenreich—continued to support the idea that the geographical variants of mankind have very deep roots in time. Ironically, these scholars were not students of modern populations but were politically liberal paleoanthropologists who strenuously defended the unity of mankind and (coincidentally) hewed more closely than any of their peers to the dictates of the Synthesis. Adopting Dobzhansky's idea that culture broadened the ecological niche of hominids to such an extent that only one kind of hominid could ever have existed at a time (albeit evidently in diverse geographical varieties), Loring Brace of the University of Michigan, one of Coon's leading critics, energetically preached the "single-species hypothesis" from the mid-1960s on. In Brace's view, which closely followed Ernst Mayr's declaration of 1950, all known hominid fossils could be attributed to one single evolving lineage. By the time that Brace wrote, the sheer amount of morphology embraced by the expanding hominid fossil record had long since caused even Mayr to relent a little. But the bandwagon gathered steam nonetheless, and by 1970 its leading proponent was Brace's Ann Arbor colleague Milford Wolpoff. It was Wolpoff who subsequently engineered the fallback position known as the "multiregional hypothesis" when, by the mid-1980s, the fossil evidence against the single-species notion had become overwhelming.

The proponents of multiregional human evolution believed (as some still do) that the fossils showed evidence for regional morphological continuity as far back in time as the earliest *Homo erectus*, a species that in some interpretations extends back to almost two million years ago. Essentially as their polycentric predecessors had done, they proposed that a single human lineage had maintained its unity by gene-flow between regional populations at their peripheries, while in the centers of their distributions local populations responded to natural selection by maintaining and refining their adaptations to native conditions. Eventually, recognizing the biological problems associated with multiple independent evolutions into the same species, they proposed to "sink" *Homo erectus* into *Homo sapiens*, so that there was no longer a species boundary to cross. In and of itself, the persistence of a single mammal species for two million years would be no surprise—indeed, it would be perfectly normal. However, the amount of morphological variety that would have to be included within the expanded *Homo sapiens* is nothing short of staggering. Moreover, it more or less beggars the imagination that steady trajectories of local adaptation should have persisted over a long period—basically, almost the entire Ice Ages—during which local environments had varied dramatically on short time scales. And many paleontologists, then as now, had difficulty in actually discerning the morphological continuities claimed by the multiregional people. Finally, both the recently much-expanded human fossil record and the molecular evidence, each of which we will shortly revisit in some detail, provide strong support for a very recent origin and diversification of our species. Nonetheless, the multiregional notion continues to find adherents in some corners of paleoanthropology.

VARIATION AND RACE

No sooner had the brouhaha among physical anthropologists over Coon's ideas died down than Richard Lewontin, an evolutionary biologist then at the University of Chicago, entered the lists. In a very influential paper in 1972 he examined the frequencies of seventeen different genes, measured in populations from around the world. His analysis focused on how the genetic variation was apportioned among and within "populations." His estimate of within-population variation is usually cited as 85 percent, meaning that most of the variation in human genes could be explained by variation within populations. Because "population" is one of those nebulous biological entities that can be defined hierarchically, he conducted his analyses at two different levels—first at the level of local populations within regions, and second at the level of regions. The former revealed that 8.5 percent of the variation could be attributed to among-population variation, and the latter approach revealed that 6.3 percent of the variation could be attributed to among-region variation. Later studies upped the percentage of among-region variation to 10 to 15 percent; but in the long run the pattern of more within-population variation than among-population variation has held firm. The notion of greater within- than between-population variation among humans was not actually new with Lewontin: the second UNESCO statement had actually made much the same point back in 1951. But Lewontin had quantified the matter, and his conclusion from the resulting figures was much more emphatic: "our perception of relatively large differences between human races and subgroups, is . . . a biased perception." What's more, "human racial classification is of no social value and is positively destructive of social and human relations. . . . [N]o justification can be offered for its continuance."

Lewontin's analysis was followed during the 1970s by several contributions that looked at other gene loci or other methods of analyzing the data. Most of these studies were in broad agreement with Lewontin, although in 1982 the mathematical geneticist Ranajit Chakraborty argued that average differences between populations and their classification on the basis of genetic markers were different questions, and that "the classification of human ethnic or racial groups remains a viable, important feature in understanding the nature and mechanism of human evolution."

The most concerted attack on Lewontin's position came much later, from the geneticist A. W. F. Edwards, who contended in 2003 that, as more genetic loci were considered, the higher became the probability that an individual could be "correctly" classified into his or her population. He thus christened the Harvard biologist's position "Lewontin's fallacy." Actually, the plant geneticist Jeffry Mitton had made the same observation in 1970, without finding that Lewontin's conclusion was fallacious. And Lewontin himself not long ago pointed out that the 85 percent within-group genetic variability figure has remained remarkably stable as studies and genetic markers have multiplied, whether you define populations on linguistic or on physical grounds. What's more, with a hugely larger and more refined database to deal with, D. J. Witherspoon and colleagues concluded in 2007 that although, armed with enough genetic information, you could assign most individuals to "their" population quite reliably, "individuals are frequently more similar to members of other populations than to members of their own."

Lewontin's Harvard colleague Stephen Jay Gould also joined the fray with his 1981 book *The Mismeasure of Man*, in which he reprised Weidenreich's excoriation of the early craniometrists. He also weighed into the conclusions of the Berkeley psychometrist Arthur Jensen, who had claimed in 1969 that the reason for an alleged failure of the Head Start programs (that had been intended to raise the IQ scores of African American children) was that intelligence was largely inherited, only some 30 percent being attributable to environmental influences. In the ensuing uproar Jensen strenuously denied that he had claimed an "'innate deficiency' of intelligence in blacks" but nonetheless declared that "there exists no scientifically satisfactory explanation for the differences between the IQ distributions in the black and white populations."

Coming closely in the wake of the prominent physicist William Shockley's resuscitation of the specter of eugenics and race, and in the midst of the epic struggle for civil rights in the United States, Jensen's claims were bound to provoke major dispute, and Gould raised once more the perennial question of whether the extremely complex attribute that is "intelligence" could appropriately be summed up in a single number. He particularly contested what he described as "two deep fallacies": the tendency to convert "abstract qualities into entities" and the propensity to order "complex variation as a gradually ascending scale." Powerful points both, but *The Bell Curve: Intelligence and Class Structure in American Life,* one of the most controversial books of the late twentieth century, was at least in part conceived as a riposte to Gould.

Published in 1994 by the Harvard psychologist Richard J. Herrnstein and the conservative political scientist Charles Murray, *The Bell Curve* was principally dedicated to two propositions. One was that intelligence as measured by IQ is an effective predictor of a host of social attributes including income and propensity to commit crimes; the other was that a dangerous gap was developing between a "cognitive elite" and the general population. But the book also contended that intelligence is highly heritable and shows significant average differences among human races as defined in U.S. society. Once more, uproar. Following some initial favorable publicity that propelled it into the bestseller bracket, *The Bell Curve* found itself pilloried on all sides. Critics included Gould, who in 1996 issued a revised version of *The Mismeasure of Man* that specifically targeted Herrnstein's and Murray's arguments. And again, what side of the argument you were on tended to be determined by your politics.

The debate still continues, fairly fruitlessly; and in February 2009 the august scientific journal *Nature* published two commentaries on the question, "Should scientists study race and IQ?" On the "no" side, the geneticist Steven Rose argued that "the categories judged relevant to the study of group differences are clearly unstable, dependent on social, cultural and political context," and that anyway "we lack the theoretical or technical tools to study them." On the other hand, Stephen Ceci and Wendy M.

Williams, whose research lies at the boundary between human development and sociology, contended that "vigorous debate has resulted in great progress in our understanding, and more breakthroughs will come—if we allow free speech in science." Both parties had defensible points of view, but each was arguing from an entirely different premise—which, of course, encapsulates the problem.

MORE STATEMENTS

If physical anthropologists—or, indeed, scientists of any stripe—were to look back and ponder a bit on the recent history of their involvement with the matter of race, you might expect them to think long and hard before muddying the waters with yet more ex cathedra pronouncements on this supremely prickly and political subject. However judiciously phrased, or however limited to strictly scientific questions, any statement you might care to make is capable of being interpreted as support for someone else's ideological framework, or as an attack on it. Yet race is undeniably a subject that touches on both biology and anthropology, and it has proven a perennially irresistible subject for biological anthropologists, who are acutely aware of the hideous travesties that have been conducted in its name.

It was, then, perhaps inevitable that by 1996 the American Association of Physical Anthropologists, unfazed by earlier debacles, should have felt obliged to enter the ring once more. In that year it issued a "Statement on Biological Aspects of Race" that attempted to clarify the old UNESCO declarations of 1949 and 1951. This well-meaning proclamation justly berated the typological racial classifications of past centuries that were based on "externally visible traits, primarily skin color, features of the face, and the shape and size of the head and body, and underlying skeleton" and pointed out that such categorizations had been used to justify the infliction of untold human misery. It also repeated the vital point that cultural and biological racial identities are often very far from the same thing.

Yet while filled with much wisdom about inherited variation both among and within modern human populations, the AAPA declaration failed to address the shortcomings of its predecessors and clearly remained confused about the historical process by which the complexly structured variation among local human populations today must have come about. The main reason for this is that, like Dobzhansky himself, the savants of the AAPA failed to complete his and their thought.

In Dobzhansky's case this was entirely forgivable, because his attempt to place race in a sensible perspective was carried out in the spirit of the reductionist Synthesis. But since the Synthesis had been formulated, much had changed in the realm of evolutionary theory. In 1972 the paleontologists Niles Eldredge and the same Stephen Jay Gould of *Mismeasure* fame had upset the evolutionary applecart by suggesting that the history of most species seemed to be not one of the slow transformation envisaged by the Synthesis but rather one of long bouts of non-change followed by abrupt replacement. And over the following couple of decades it came to be widely acknowledged that, in a climatically and environmentally capricious world, the evolutionary process had necessarily consisted of much more than gradual fine-tuning of lineages via gene frequency change. Change that we see as linear in retrospect may well actually occur in the context of simultaneous differentiation of local populations of the same widespread species, as they adapt to their local circumstances or as they reflect peculiarities that were present in their founders. Yet only when speciation intervenes—when an impediment to interbreeding arises—do those populations become independent entities on the evolutionary scene.

It was thus fair enough for the AAPA to say that "pure races do not exist in the human species today," but less accurate for it to continue that "nor is there any evidence that they have ever existed in the past." As far as we know, in a complex mammal such as *Homo sapiens* significant population differentiation can only occur in isolation (as we've pointed out, the Weidenreich model and its derivatives are like having your cake and eating it), and small, isolated populations were the rule over the Ice Age history of mankind. If an isolated population were over some extended period to acquire unique local evolutionary novelties, then that population could indeed be said to be a "pure" race—until it began to lose its unusual features by merging with contiguous populations of its species, when reunited by climate change or some other factor. As one of us (Ian) said in an article published in 2004, "it is as if in order to refute the reality or utility of the notion of race in the world today, one must deny that there was any geographically-based differentiation among earlier human populations by insisting that, somehow, human variability has always been patterned in the same general manner."

The mindset we are lamenting here makes it all the more important to complete Dobzhansky's thought. As he pointed out, "pure races" do not now exist. But, in order to have achieved the pattern we see in humankind today, fairly clearly differentiated populations must have existed in the past. Certainly, at a relatively recent stage population differentiation must have been more clearly marked than at present. The key to the current confusion is that, however varied they may seem to the eye (and the human eye is practiced indeed at discerning even tiny differences), all human beings nonetheless belong to the same species, *Homo sapiens*. Individual cases of infertility aside, all humans today are completely interfertile, and the vigorous interbreeding that is going on, now that population contact is the rule, is blurring the physical distinctions that had, long ago, developed in small isolated populations. It may very well be that the complexities of the murky human psyche will conspire with cultural influences to slow this process down in *Homo sapiens,* but there is no doubt that it is inexorably under way.

So the crucial point here is that *inside* any species, *Homo sapiens* included, *reticulation* is possible, meaning that genetic lineages will both appear and disappear—in the latter case via combining with others as well as by extinction. This fact of reticulation explains why a recent attempt by the philosopher Robin Andreasen to challenge the idea that biological races don't exist simply does not work. Andreasen argues that "cladistic races" can indeed be recognized because of the branching of lineages that takes place within species. Races are simply a product of this branching process. This is true as far as it goes; but it's only part of the story. For *within* a species those lineages may also freely recombine, as they are doing in ours today. The resulting weblike pattern of relationships is very different from the "dichotomous" pattern *among* species that the "cladistic" approach to recognizing evolutionary relationships was developed to clarify.

The bottom line here is, then, that within *any* species, two basic genetic processes are possible: *differentiation* and *reintegration*. In the conditions that reigned during the last Ice Age, human beings lived in small populations that were widely scattered over the continents and that were extremely thin on the ground. And it was under those circumstances that the ancestors of today's major groups clearly differentiated, in effective isolation from each other. But since the giant polar icecaps began to retreat, some twelve thousand years ago, there have been huge human population expansions, resulting in the mingling and reintegration that is now the rule.

The alternating operation of this dual process is, above all else, why it is simultaneously possible intuitively to recognize races and hopeless to define them biologically or to draw clear lines among them. Races as we perceive them at any point in time are not in the least static: they are actually part of an ongoing dynamic. It is certainly ludicrous from all points of view that the United States government should insist on classifying Ian's stepdaughter in a different race from her mother, and she is certainly right to be unhappy at having regularly been required to shoehorn herself into an inappropriate pigeonhole. The ability to check more than one box in the latest census came as a welcome relief—but what about the next generation, and the one following? The reason for the logical and practical dilemmas so clearly posed by a rigid classification of races is, of course, that this is most emphatically not at root a biological issue: it is a cultural and historical one. Because of the rampant tendency for populations of the highly mobile contemporary *Homo sapiens* to intermix, together with our apparent ingrained need to identify with much smaller units than the species as a whole, races as we understand them today are above all sociocultural constructs, the products of complex ethnic, cultural, and geographical histories. Typically, such constructs are laboriously built over time, according to a host of political and demographic vagaries, as boundaries are defined and redefined by those in power. But in the end the outcomes are entirely arbitrary. Take, for example, the latest U.S. census, which lists several kinds of Asians but lumps all Europeans and Africans into single categories. No biology here.

Nell Irvin Painter has elegantly provided an example of this sociocultural process in her recent book *The History of White People*, in which she traces the construction of the concept of "whiteness." She finds the beginnings of the recent idea of a "white race" in the writings of eighteenth-century German scholars (although it is noteworthy that in the United States the notion was already current a century before and was enshrined in law as early as 1691 in the colony of Virginia, where a decree prohibited "abominable mixture" through marriage between "whatsoever English or other white man" and "Negro, mulatto or Indian"). Foremost among the German scholars Painter cites was Blumenbach, the author of the white-skinned "Caucasian" race, a term that still pervades our vocabulary today, having come back into vogue after later iterations had been discredited by Nazi atrocities. Painter points out that such American luminaries as Thomas Jefferson and later Ralph Waldo Emerson reworked the Caucasian ideal into one of "Saxon" superiority, a category terminologically succeeded by labels shifting to Teutonic, Nordic, Aryan, and then to white/Anglo—invariably in support of existing social stratifications, power structures, and perceived advantages. Superiority was illustrated in different ways at different times and sometimes simultaneously; it was implicit, for example, in the prewar eugenics campaigns and was made more explicit in the IQ debates that continued into the late years of the twentieth century. The upshot of such histories as this is that our practical notions of race are born in our heads or are acquired by them. And they seem to derive not only from powerful political expediencies but, on a deeper and more enduring level, from the innate human need not only to *belong,* but to represent this belonging intellectually, as part of some arbitrary classification.

This is, of course, why we started this chapter with the observation that classifying everything (not just ourselves) is an inherent part of the human way of doing business. The fact that our notions of races are arbitrary products of our cognitive processes rather than direct translations of biological reality is, above all, why it is unwise for the biologists of the AAPA to take an official stand on the matter, as if the existence or the nonexistence of human races requires a special biological explanation that physical anthropologists are uniquely equipped to provide. There really isn't anything special to explain. In evolutionary terms, both the differentiation and reintegration of populations within species are entirely routine phenomena, and in this regard, at least, *Homo sapiens* is just another successful widespread mammal species.

GENOMICS AND MEDICINE

There is one area, though, within which the classification of individual humans into races might conceivably have some utility. This is in medicine, where belonging to a particular population might constitute a statistical risk factor for a specific disease. Because if you know you are at risk, you can take precautions. Black Americans, for example, are said to be at unusually high risk of developing hypertension, while some Amerindians are particularly prone to diabetes. The problem is that such groups are typically defined culturally so that, as we will discuss later, group risk as determined by biology may well not apply to a great many individuals. We are far from the only ones to have noted this, and in his recent book *The Emperor's New Clothes*, Joseph L. Graves Jr. has well clarified what he calls the "Race and Disease Fallacy." What's more, as we will soon see, the fact that people will be able in the near future to carry around an affordable characterization of their entire genetic makeup on something the size of a credit card means that we will soon be able to directly know our risk for all sorts of conditions as individuals, rather than indirectly as members of a group that may or may not be relevant. Or then again, maybe we would rather not know.

Still, it's unarguable that the visual stimuli to our perceptions of race have some basis in heredity. And this fact seems to have made the question of race an irresistible one to geneticists and most particularly to students of the burgeoning science of genomics. Genomics has proven an excellent tool for deciphering the history of human movement throughout the world following the exodus of our species from Africa some sixty thousand years ago, and it has spurred a widespread interest in the molecular tracing of "ancestry," a word that, unencumbered by the baggage accompanying the term "race," is nowadays often substituted for it. Sadly, however, race and ancestry are far from the same thing. "Ancestry" has a clear meaning in the context of tracing the familial descent of individual human beings, and it has an equally clear meaning when applied to groups of species that, like individuals, are discrete entities. But when it is applied to populations within species, reticulation renders this term merely confusing. An individual can derive his or her ancestry from a number of populations possessing different geographical origins and histories of intermixing. The large literature now accumulating on genomics and race has thus not succeeded in clarifying the race issue much, as we will see in later chapters. We are still left with the difficulty that "races," with their complex geographical and cultural histories, defy any operational criteria for delineation.

It is this reality which, among many other things, undermines the many efforts that have been made to demonstrate the efficacy (or its lack) of a whole range of drugs in human subpopulations with diverse assumed geographical origins. In order to test such propositions you have first to sort your sample of subjects into discrete groups. This grouping is usually done by the subjects themselves, in a process of self-identification based on an established classification. And the group a particular individual identifies with is infinitely more a matter of culture and history than it is of biology—casting a huge amount of doubt on the biological validity of the associations arrived at.

Yet government policies in the United States are having the presumably unintended effect of entrenching these unsatisfactory categories. For example, congressionally mandated policy at the National Institutes of Health, the major funder of biomedical research in the United States, currently forces clinical scientists to choose their samples of research subjects along ethnic/racial lines. Grants will only be awarded if "members of minority groups and their subpopulations" are included in the research design—regardless of whether such categorization is of any relevance whatever to the research at hand. More than one minority scholar has lamented that this official "emphasis on racial differences" helps to reinforce the notion that the biological differences among human populations are somehow significant. Clearly, our rapidly increasing knowledge of the molecular genetics underlying human individual and population differences has not decreased the need for caution when establishing political policy relating to this issue.

We will not further belabor the current state of affairs here, because we will look at these issues in much more detail in chapters 2 and 5. But we would still like to make one final point before concluding this historical survey. And we think it is a very significant one, namely that none of what we've just said means that, as the third millennium begins, we should be totally ignoring human population diversity. Classifiable or not, it's a fact of life, and the molecular geneticist Bruce T. Lahn and the economist Lanny Ebenstein have recently urged in the pages of *Nature* that it should be "embraced and celebrated as one of humanity's chief assets." One reason for this declaration may have been that Lahn had, in the words of *Science* magazine's Michael Balter, run into "heavy fire" following his research group's finding that a couple of gene variants, both of them involved (along with many others) in brain size determination, not only occurred at much higher frequency in Eurasians than in Africans but conferred an adaptive advantage. Further work rapidly falsified the implicit suggestion that the variants were associated with increased intelligence, and as a first-class scientist Lahn rapidly distanced himself from this implication.

Still, presumably somewhat in shock from the barrage of criticism, in their *Nature* piece Lahn and Ebenstein also deplored the political correctness of the "biological egalitarianism" that, in over-response to hideous historical injustice, had come to pervade discussion of variety within our species *Homo sapiens*. Taking something of a middle road, they pointed to the possibility that scientific data might "ultimately demonstrate that genetically based biological variation exists at non-trivial levels not only among individuals but also among groups" and suggested that "the scientific community and society at large are ill-prepared for such a possibility." They concluded with the ringing declaration that "Equality of opportunity and respect for human dignity should be humankind's common aspirations, notwithstanding human differences no matter how big or small. . . . [H]uman genetic diversity as a whole, including group diversity, greatly enriches our species."

From whatever perspective you might care to interpret the first part of this message, an evolutionary biologist would certainly resonate to the idea that, to the extent that they amplify human diversity, group differences provide a potential evolutionary advantage to our species. Even a cursory examination of major patterns in the history of life is enough to demonstrate that it is in everybody's interest that our species should vary: the fossil record shows clearly that narrow specialists have much higher extinction rates than generalists do, and apparently it is rarely wise for a species to have all of its evolutionary eggs in one basket. Monocultures are notoriously fragile, as witness the apparent impending extinction of the familiar Cavendish banana due to the fact that, worldwide, virtually all cultivated bananas are genetically identical and equally susceptible to the ravages of the "Panama disease" currently raging among them.

Lahn and Ebenstein emphasize that their perspective by no means ignores the effect of environmental influences on human variation, that it does not diminish the importance of within-group variety, and that any potential downsides to group genetic diversity do not detract from its overall benefit to our species. And they declare with some passion that "humanity has been and will be stronger, not despite our differences but because of them." Where the argument will go from there—for without any doubt, debate and even invective there will be—it is tough to predict. But since almost everything possible has already been said; since it is easy to be seen as protesting too much; and since to take virtually any position on this issue will only fan the flames of sociopolitical argument, biologists might find it preferable to stay out of the picture. Providing facts is one thing; deploying them is very much another.

THE COMMON THREAD

The principal common thread running throughout this chapter is the uncertainty that continues to accompany just what is signified by the word "race," a term that we would be much better off without but that has so far resisted expulsion from our vocabularies. Originally introduced as vague equivalent to the equally elusive "species," the word has since become laden with layer upon layer of confusion and has acquired a multitude of meanings. One way of understanding what has happened to "race" as a biological concept might be by comparing it to the "telephone game" that many readers will recall having played as children. A phrase is whispered into one person's ear at the start, and then this person whispers it to the next, and to the next, and so on, until the last person hears the phrase and blurts it out. Inevitably, no matter what the last person thinks he or she heard, it will have no resemblance whatsoever to the initial phrase. In the following chapters we will attempt to play this telephone game backwards, to try to get at the root of what race has meant to various authors and audiences, and at what it really means in a biological context. Along the way, we will need to discuss evolutionary process in a bit more detail; and since "race" can only have biological meaning in a taxonomic context (after all, the word itself is a taxonomic epithet, albeit an informal one), we will also need to look briefly at taxonomic principles and procedures: the ways in which we classify living things and understand their diversity. Finally, to complete the picture, we will also have to discuss how species are delimited and how the notions of "race" and "subspecies" are involved in this technical process.

CHAPTER 2

SPECIES, PATTERNS, AND EVOLUTION

I N FEBRUARY of 2010, the beautiful southern U.S. city of Atlanta, Georgia, was
besieged by a billboard campaign that claimed that black children were becoming "an endangered species." Antiabortion proponents sponsored the installation
of over seventy-five of these sensationalist billboards across Atlanta, mostly in
African American neighborhoods. Regardless of the politics involved, scientists were
shocked at the blatant misuse of the terms "endangered" and "species" in this campaign. Because there is so much confusion regarding the terms "species" and "race"
with respect to humans, we feel it important to set the record straight here. No human
population, or social group, is a species, and it is critically important that everyone understand how biologists use the terms at issue here—especially since a complete understanding of the concepts they embody lies at the heart our argument for the "burning"
of the term "race" as a biological epithet among humans.

From everything we've said so far, it will also be evident that the key to understanding the way human diversity is patterned lies in a proper understanding of the
evolutionary process which produced our species *Homo sapiens*. This is why we will
spend the next few pages discussing this widely misunderstood process and how we go
about understanding both the process itself and the historical patterns it has produced.
After that, we will be in a position to tackle the human fossil record that documents
exactly how it was that human beings came to be as they are. Because humans are
mammals, and have thus been subject to all of those influences that affect the evolutionary history of any successful mammalian group, we think it's important that the
reader should have some acquaintance with the systematic context within which all
mammals have to be understood. In this chapter we may provide more detail on this

than some readers may care to acquire, in which case we encourage you to skim the following sections on your way to our more focused discussion of human races. But we do believe that you will be rewarded by a closer reading.

THAT FOUR-LETTER WORD

The human diversity we see around us today is simply one single, if extreme, example among millions of evolutionary experiments in the almost four-billion-year history of life on Earth. The results of other such experiments are everywhere: just look around you. Even if you are indoors right now you are surrounded by hundreds of other expressions of evolutionary history. If you are inside, perhaps you see your cat or dog, an obvious example. Through the window you might see some trees and other plants outside in the garden; again, very obvious. If you were able to look closer, you might even see several kinds of small wormlike creatures called nematodes crawling around in the soil, or on and in the plants themselves. Back indoors, you might look at your carpet closely. Lurking in the pile would be mites, many of them: depending on how often you vacuum, there might be as many as one hundred thousand in a single square yard of your carpet. If you could look even closer, say at a dirty tabletop, you would see a huge variety of bacteria; a mere spoonful of dirt from that backyard outside would yield hundreds of different kinds. Then again, the air you breathe is billowing with huge numbers of living things—bacteria, fungi, and even the reproductive cells of eukaryotic organisms (pollen from plants). In fact, to assess the diversity of the results of evolution you need not go any further than your very own body or even your mouth. It has been estimated that the hundreds of kinds of bacterial cells living in and upon you outnumber your own cells by ten to one. In your mouth alone, even if you brush your teeth regularly, there are over four hundred kinds of bacteria resident.

These are just some of the thousands of different kinds of living things we encounter each day. But exactly what do we mean by "different kinds of living things"? In order to understand our place as humans in the grand evolutionary experiment, and to make sense of how we view ourselves, we need to understand how living things become discrete enough to be recognized as individual "kinds." We will never understand why we see the diversity of people that we do around us, organized in the way we perceive, if we do not understand the evolutionary processes that produced ourselves and the millions of other living organisms on our planet. This necessity becomes even greater when we consider that only 0.01 percent of the life forms that have ever existed on our planet are still living today. This means that 99.99 percent of the kinds of organisms that have ever populated the Earth have gone extinct, skewing perceptions based on our immediate experience. So to understand the larger picture of which we are part, we need to delve a bit into how living beings evolve and diverge enough to be recognized as different and how they go extinct. Most importantly, we have to appreciate that the evolutionary processes and interactions that take place *within* kinds of organisms are qualitatively different from those that operate *among* kinds of living things.

The astute reader will notice that we have used the word "species" only once in the preceding paragraphs. This is because we feel this word is all too often completely and utterly misunderstood. Now, it might seem strange that two natural historians should suggest that the single most important word in the language of evolutionary biology is misunderstood. After all the word "species" occurs right there in the title of the single most important work in the science of evolution, Charles Darwin's *On the Origin of Species by Means of Natural Selection*. His title wasn't *On the Origin of the Orders of Organisms* or *On the Origin of Kingdoms of Organisms* or even *On the Origin of Small Groups of Organisms*. On the other hand, while the title implies a focus on species, it has been suggested that Darwin really sidestepped the species problem. In fact Ernst Mayr, one of the authors of the New Synthesis, suggested that the *Origin* was "misnamed" and was not in fact a "treatise on the origin of species." Many authors since have sided with this view of Darwin's *Origin*.

We contend, however, that Darwin's use of the word "species" in his title was no accident. This is because Darwin pretty much got it right that the focus of evolutionary studies should be on how living things diverge and differentiate. Darwin is most famous for having presented overwhelming evidence (he himself called his book "one long argument") for the existence of evolution; for having posited a mechanism (natural selection) by which evolution could proceed; and for changing the typological approach of natural historians of the time to a focus on populations and processes. For some reason, though, he is less well known for a fourth idea that he presented equally clearly in the *Origin*, namely, the principle of divergence. In a chapter entitled "Mutual Affinities of Organic Beings," he presented his famous genealogical tree (the only figure in the entire book), declaring that "thus, as I believe, species are multiplied, and genera are formed." This tree is a graphic representation of lineages diverging.

Darwin most certainly understood the critical importance of the "principle of divergence" to an understanding of evolution and indeed emphasized this in a letter

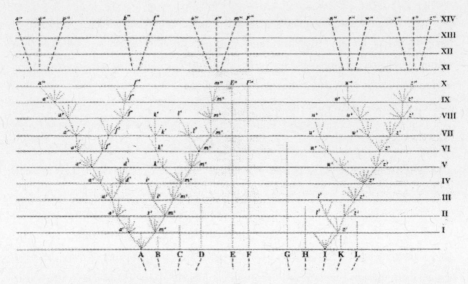

FIGURE 1. *The only figure from Darwin's* On the Origin of Species. *Darwin intended this diagram to represent how the principle of divergence operates. The Roman numerals on the right indicate generations, and the various uppercase letters represent the ancestors of the various lineages that diverged during the genealogy of this hypothetical group of organisms. The lowercase letters represent subsequent intermediate ancestors in the genealogy. Lowercase letters with two prime marks (") indicate terminal taxa or individuals.*

dated June 8, 1858, a few months before the book was published, that he sent to his friend and colleague J. D. Hooker. He would, he wrote, "discuss the 'Principle of Divergence,' which along with 'Natural Selection,' is the keystone of my book; and I have very great confidence it is sound." Criticisms that Darwin neither completely nailed the species problem, nor provided a workable definition of species, may have some truth to them. But Darwin did clearly understand the importance of divergence, and he strongly implied that species were an integral part of this principle. Indeed, understanding the nature of species had been the goal of much of natural history up to Darwin's time and has been since. From the beginning, natural historians have known that it is essential to understand the diverse array of living things around us. And understanding speciation—the process by which discrete new kinds of organisms come about—is the single best way to understand this diversity.

Still, "species" is a loaded word, and to see how loaded it really is, we need look no further than the subtitle of Darwin's book, the full title of which is *On the Origin of Species by Means of Natural Selection, or The Preservation of Favored Races in the Struggle for Life*. In his subtitle Darwin substitutes the word "race" for "species" and "struggle for life" for "natural selection." The substitution of "struggle for life" for "natural selection" makes a certain amount of intuitive sense. But interchanging the word "race" for "species," even casually, can cause immense problems. This is why we are devoting this chapter to unraveling the confusion surrounding this simple four-letter word. Similarly, since "race" is often equated with "subspecies," a semiformal rank in the hierarchical classification of living things, it is also helpful to know the rules by which organisms are classified. Below we look both at how evolution occurs at the level of species and at how its products are classified.

RIPE FOR BURNING

As we mentioned earlier, modern evolutionary biology cut its teeth in the first half of the twentieth century with the development of the New Evolutionary Synthesis. In the process of developing the Synthesis, very discrete subdisciplinary barriers were erected in evolutionary biology. But before we tackle these subdisciplines, we have to define just what evolution is. The simplest and most elegant definition of evolution was developed after the rediscovery of Mendel's laws of inheritance (the foundation of the modern science of genetics and ultimately of the New Synthesis) at the turn of the twentieth century. Without those laws, our modern perception of the evolutionary process would almost certainly be quite different. The framers of the Synthesis placed great emphasis on an empirical and theoretical understanding of genes, the units of heredity. That is, they emphasized the dynamics of how different forms (alleles) of the same gene function in populations. Within this focus, the definition of evolution came to be simply "change in allele frequency with time."

It is this definition that led to the independent development of those discrete subdisciplines in evolutionary biology. Why? Well, first and foremost it led directly to the science known as population genetics. By following the behavior of genes in populations, this branch of biology directly addresses what has come to be known as *microevolution*, or evolution at the population level. And while from the beginning it included scientists who study the process of evolution in natural populations, it more or less excluded those interested in *macroevolution*, the area of evolutionary biology that studies large-scale changes in organisms, interested as it is in interpreting the larger patterns that the evolutionary process has wrought. What's more, while microevolution and macroevolution together cover a lot of evolutionary ground, something is still missing: namely, how species, the major currency of evolution, are formed.

The renowned biologist John Maynard Smith once made a comment about an important evolutionary term—"homology"—that had long confused and polarized biologists. He suggested that this widely used term was "ripe for burning." Well, he certainly had a point, but scientists have not burned it yet, and they probably never will, because when they use it everyone more or less knows what they mean. Biologists can have reasonable discussions about the concept of homology, even if they disagree about exactly what it is, because it is the details that usually cause the problems, not the fundamental reality. So also is it with the terms "species" and "race." As words that have for generations mystified, polarized, and downright confused biologists, these are terms that might appear more than any others in biology to be ripe for hanging, burning, and burying. And indeed, part of the thesis of this book is that one of them—race—could be expunged with huge advantage from the vocabulary of biology, at least as it pertains to human beings. On the other hand, there is something about the term "species" that prevents it from being ignored and placed into the bin of useless scientific terms alongside phlogiston, cold fusion, and pangenesis. And the reason is, of course, that it still works. When a biologist uses the term "species," other biologists still have a good idea of what is being discussed, and even if they disagree about its precise definition, they can still communicate using both this word and the concept behind it.

Species are also at the heart of our current classification schemes, as we will see later on in this chapter. In contrast, "race" does not have this biological utility. It's doubtful that it did even when taxonomy was first being developed as a science. And there is little prospect that it ever will, hence its true ripeness for burning. But before we light the fire, let's look a little more closely at the vocabulary of terms that are indeed useful in understanding the evolutionary process at this crucial level that stands between microevolution and macroevolution.

TOO MUCH AND TOO LITTLE SEX

Perhaps the best recent definition of species came from Ernst Mayr, who encapsulated it in the 1940s, basically re-rendering the seventeenth-century suggestion of John Ray. Mayr's thumbnail definition, that species are "groups of actively or potentially interbreeding individuals," has become an enduring concept in modern biology. Nonetheless, Mayr's "biological species concept" has regularly been pilloried by opponents, and numerous alternative definitions have been proposed. Basically, adversaries point to three major shortcomings of Mayr's definition. And since interbreeding is the linchpin of the definition, we can explain the first two shortcomings with sex—either "too little of it" or "too much of it."

Most of the organisms on this planet are microbes that don't have sex. These organisms, best represented by bacteria, reproduce by cloning themselves and hence reproduce asexually. Bacteria form well-defined lineages, and, according to the microbiologists who study them, they make good species. In fact, they speciate so well and so efficiently that scientists estimate there are at least two million living oceanic species and at least four million living terrestrial species of bacteria (and probably millions more). If one took the biological species concept to its extreme, each lineage of bacteria would logically be its own species. However, microbiologists resist going this far and clump closely related lineages into single species. They do this based on how similar the different closely related bacterial lineages are to each other and use what seem to be reasonable cutoff points as species boundaries. This process creates an interesting situation where bacteria have "too little sex" (actually no sex) to be covered under the biological species concept.

Believe it or not, there is also such a thing as "too much sex." This happens when two species that are generally recognized as exemplary reproductively isolated entities actually "sneak one in." An excellent example is found in Haleakala crater in the Hawaiian Island chain. One of the more bizarre earthly habitats, it is much like being on the moon, except for occasional stands of plants and the fact that there is air there. Appropriately, one of the predominant plant species there is an exceedingly odd-looking plant of the family Asteraceae (sunflowers are in the same family) called the silversword. As their name implies these plants are silvery concoctions, with sword-like leaves emanating from a central shaft. They are classified in their own genus *Argyroxiphium*. They grow to pretty good heights, sometimes to five or six feet and taller; and when the sun shines on them in the crater they are a sight to behold, glistening and gleaming like extraterrestrials standing at attention. When you look at them closely, they loudly proclaim, "look at me, I am a member of a good species," because they appear so different from other plants on the other Hawaiian Islands that are presumed to be closely related to them.

What is really bizarre about these plants, though, is that they can interbreed with plants in another genus—not just another species, but another genus. The botanists who work on them are convincing when they classify the silverswords from the vol-canoes of Hawaii into two genera (*Argyroxiphium*, endemic to the islands of Maui and Hawaii, and *Wilkesia* on Kauai), while putting related species also from the Hawaiian chain in the third genus *Debautia*. After all, the first two genera look very dissimilar from the third and have different habitats and ecological requirements: *Debautia* is a low-profile species that has none of the spectacular alien look to it that *Argyroxiphium* in the Haleakela Crater has. Not only are these plants good species, but they are good genera; yet they have "too much sex" to fall under the definition of the biological spe-cies concept because they can "do it" and, more importantly, produce offspring. Yes, they are hugely differentiated physically, but no, they are not isolated reproductively.

The problems of the bacteria and the silverswords represent a failure of the bio-logical species concept to be theoretically encompassing. The third shortcoming of the concept often cited by opponents is more a matter of practicality. To be used, any definition in science has to be useful, usable in the field as it were. The problem with Mayr's concept is that in order to use it we must be able to determine if individu-als from different populations and different species are actually "doing it" and if so, whether or not they are doing it to any significant effect. As young scientists, both of us sat in the field waiting to see individuals from the species we worked on "do it." We spent a lot of time as peeping toms, because how your study organism "does it," and who they do it with, is an important part of understanding the identity of the species you are studying. For one of us (Ian), the incredibly short mating season of lemurs was one of the keys to recognizing whether related forms should or should not be assigned to the same species. For the other (Rob), the task was to watch tiny nondescript fruit flies congregate on limbs in the rainforests of Hawaii, hoping for them to literally "do a little dance, make a little love, and get down tonight," because males in this group actually do a courtship dance that entices the female to copulate. Hundreds of field hours of observing these flies ended in frustration, with no sightings of dances, let alone copulations amongst the species. It is a whole lot easier to observe two barnyard animals copulate than it is to observe animals in nature doing the same thing. Even more difficult is observing how plants interbreed, because they don't require physical contact to do it. Often, indeed, they allow other organisms to implement the delivery of gametes from male to female. So there is a huge operational problem when applying Mayr's concept. And there are lots of opportunities for the organisms you're observing to "cheat" and stray from their conspecific mates.

Still, while in science we want to have ideas and definitions that we can use in practice, another quality that makes a definition scientific is that it is objective. Any introduction of subjectivity into the definition allows a slide down a slippery slope toward scientific uselessness. And one thing that can be said for Mayr's concept of species is that it is objective in the extreme. If you can pin down whether things are interbreeding or not, then you can unambiguously apply the definition. Whoever does the analysis, if you have the right information on interbreeding, everyone will come to the same conclusion. That makes for good science. Which leaves us with a conundrum: Mayr's definition is not always operational, and it is slightly theoretically unsound, but it is about as objective as any definition of species can be. Should we ignore its theoretical quirks and its operational deficiencies for the advantage of objectivity? Luckily, there are other criteria for species recognition that we can fall back on that are both operational and objective and that in many ways fit with Mayr's definition.

THE LINE OF DEATH

At last count there were over twenty-five species concepts or definitions in the current scientific literature. It seems that some of the ideas about how to define species have actually speciated themselves! Many of the species definitions that have been suggested since Mayr's use the same basic idea but change the criteria for recognizing reproductive isolation. Whereas Mayr's criterion depends on whether or not organisms are having reproductively successful sex, these other approaches attempt to use proxies for recognizing reproductive isolation. They are thus technically not new concepts but rather offer new criteria for the recognition of biological species as defined by Mayr.

The approach adopted by Mayr advocates using the patterns seen in nature as criteria to infer whether a process of species divergence has occurred. Some of the newer ideas about species formation that have been developed recently take a novel tack on the subject by turning it upside down. They attempt to use process-oriented criteria to take into consideration the "too little sex, too much sex" problems that Mayr's concept involves. These process-oriented criteria include genetics and demographics, two factors important in divergence of populations as determined by modern population genetics. We consider the incorporation of these considerations enough of a departure from Mayr's biological species concept to warrant regarding it as a new species concept. Known as the "cohesion species concept" (and formally defining species as "the most inclusive group of organisms having the potential for genetic and/or demographic exchangeability"), this view of species was developed by the biologist Alan Templeton. Whether or not the cohesion concept is accurate and useful in speciation studies is debatable, but it is indeed a new kind of idea about how species can be defined and detected.

Another approach has been termed the "phylogenetic species concept." In this case, however, it is debatable whether or not this is really a full concept. Perhaps it is more properly seen as the most concise set of criteria yet developed for recognizing biological species in practice. The phylogenetic species concept suggests that "species are the least reducible diagnosable units in nature." This means simply that the unit possesses a feature (or several) that unambiguously diagnoses the group to the exclusion of all other potential groups. If you cannot reduce the group further, then it becomes the least reducible group, and hence the species. Especially given DNA sequencing technology, this definition is highly operational, and it is also highly objective. Simple algorithms can be written to accommodate the criteria established for the phylogenetic species concept. But is this really a concept? Is it fundamentally different from Mayr's biological species idea?

We think not. The phylogenetic species concept arose from the writings of Willi Hennig, a German entomologist who was interested in systematizing flies. He pointed out, as Darwin did, that organisms evolve by the "principle of divergence." This principle suggests that organisms evolve by bifurcation from common ancestors, to form new species. In one of the most reproduced figures in systematics (Darwin's genealogical tree from the *Origin*, its only illustration, is the single most reproduced), known informally as "Hennig's Figure 6," he drew out how a single ancestral population would diverge into two. At some point, a genealogical disconnection occurs, and the two new lineages evolve until they are diagnosably distinct and have hence become new species. Obviously, this requires a break in the connection between the ancestral and descendant populations. And what might that break consist of? Simply put, it's a lack of interbreeding between the two daughter populations. Sound familiar? This very familiarity is why we suggest that the phylogenetic species idea is not really an independent concept but rather a set of hard-nosed criteria that implement the biological species concept of Mayr.

Hennig's figure 6 is actually one of the more interesting figures in all of modern biology. We would place it up there in influence with Watson and Crick's famous double helix figure from their 1953 *Nature* paper or even with Gary Larson's "the real

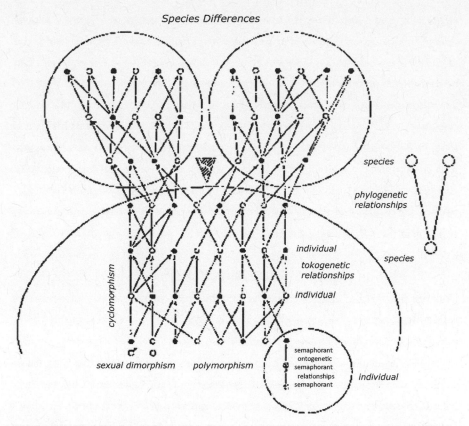

FIGURE 2. *Hennig's famous "figure 6." The small circles represent individuals in a genealogical relationship. Each individual above the bottom line has two arrows pointing to it and leading back to its parents. The gray triangle indicates a geographic or other barrier that separates the initial single interbreeding or "reticulating" population into two new species, with no genetic crosstalk. The two large circles at the top encircle the two new species.*

reason why dinosaurs became extinct." Besides clarifying how characters can be used as proxies for reproductive isolation, this figure also makes an amazingly clear statement about how we need to view the speciation process. It shows that there is a point where speciation occurs and that this point is where the tree-like structure of any diagram of relationships among species arises.

Below this critical point, while trees *can* be drawn for individuals, such trees mean little with respect to the true relationships of the individuals concerned. This is because, below the point of speciation, organisms are continually cross-talking genetically (interbreeding), swapping genes and merging lineages so much so that any

attempt to draw a tree for these individuals results in a net, or a tangle. Some systematists have rather alarmingly called the boundary the "line of death." Below the
line of death, where species do not exist, tree-building methods do not represent true
relationships of individuals. Here allele frequencies in populations are the currency of
analysis, and population genetics reigns supreme. Above the line of death, where species are detectable, trees represent relationships among species, diagnostic characters
for taxa are the currency of analysis, and systematics is king. We will see later that some
molecules have been used to draw trees for individuals. These actually represent valid
exceptions to the rule just adumbrated, but trees based on individuals relate to only two
very special cases out of the twenty-four thousand that exist in our genomes, and we
will look at them in detail a little later.

TOP DOWN AND BOTTOM UP

This brings us to the matter of how we actually look at the events, processes, and
patterns that occur at and around the "line of death." The methodology for studying
populations at the microevolutionary end of the spectrum is well entrenched in evolutionary biology and has been around for over a century, since the rebirth of the field of
genetics. With its reliance on frequencies and demographics, the study of population
genetics has grown into an incredibly broad and interesting area of research. Systematics, the science that studies evolutionary phenomena above the line of death, has an
even longer history, and while its present incarnation is a bit younger than population
genetics is, the use of phylogenetic trees throughout its existence as a discipline is a
significant common thread between the two disciplines.

There are two significant issues here. The first is the infusion of tree-building
methods into population genetic studies below the line of death—which is, of course,
what we're interested in when we are trying to reconstruct the history of human populations around the world. The second involves the practicalities of studying speciation,
given that both systematics and population genetics have entrenched methodologies.

The first issue has been helped along by the development of what is called "coalescent theory." The classical phylogenetic approach attempted to reconstruct history
in a "bottom-up" fashion, by going *from* ancestors *to* descendants. The developers of
coalescence realized that if one looked backwards at a geneaology, *from* descendants
to ancestors in a "top-down" perspective, an interesting thing happens: events occur
where the descendants coalesce into an ancestor. These coalescent events turn out to

be highly informative, and, using current population genetics methodologies, interesting techniques can be applied to unravel important historical events. Such events might include the range expansion of a species, the role of migration in how a genealogy unfolded, the estimation of the time to coalescence, and other factors important in the divergence of a species.

This is an incredibly powerful approach that works well in practice, and it has a strong statistical and mathematical framework. What's more, it has been used to good effect in the reconstruction of the biogeographical history of *Homo sapiens*. But we must keep in mind that, when coalescence is used, it is used on single genes with histories independent of other genes. Some newer coalescent methods are used to understand the linkage of genes to each other, but the basic approach still focuses on single-gene genealogies. Perhaps in the future a coalescent model for whole genomes will be developed, but for now, all of the inferences made using the coalescent approach use single-gene genealogies as a starting point. And while we think this is a valid approach for understanding the coalescence of the genes under inspection, we also believe that extending the results of one coalescence analysis to make statements about the evolution of individuals should be done with caution. One gene will tell a particular story—which might well be completely contradicted by the story told by the next gene.

The second issue concerns the methods of systematics and population genetics and how these impact the study of speciation. For the most part, systematics uses patterns gleaned from the present to interpret the past. Fossil evidence gives us a great window on the past from the standpoint of past anatomy. Study of ancient DNA does something similar, in giving us a window on the past from a genetic viewpoint. The limitation here is that the application of systematic techniques to the study of past speciation can only produce inferences about patterns in the speciation process. But even so, the top-down approach can reveal patterns that tell us a lot about speciation and its end result—species.

Two examples of the impact of this approach on evolutionary biology will show how important the top-down approach has been. The first concerns the early work of Niles Eldredge and Stephen Gould, originators of the "punctuated equilibrium" notion we mentioned earlier. Their work resulted in a whole new way of looking at how speciation could occur. As we've seen, for about a century evolutionary biologists had been convinced that evolution proceeds in a gradual, slow, "imperceptible" (Darwin's

word) fashion. In fact, Gould eloquently wrote later about this entrenchment of gradu-
alism when detailing Darwin's preoccupation with worms and "vegetal molds" and the
gradual changes they wrought in landscapes. What Eldredge and Gould discovered
is that, at least in some organisms, evolution has been episodic rather than smoothly
gradual. Using only patterns they observed from fossils, they proposed that some species
showed long periods of no change ("stasis") followed by brief bursts of differentiation.
Then the cycle would repeat itself. This pattern-based proposal resulted in a major shift
in how evolutionists viewed the speciation process.

The second major pattern-based advance in evolutionary biology involves taxonomy (see text box, p. 85, for a definition of taxonomy as it relates to systematics),
an important discipline for understanding natural diversity. Its historically typological,
pattern-based approach made it a key player in the new systematics, since delimiting
species in this way is a top-down endeavor that is highly dependent on the patterns we
observe in nature. We will return to taxonomy in more detail in a page or two.

Speciation can also be studied from the perspective of population genetics, using a bottom-up approach that may more directly allow the discovery of processes that
are involved in speciation. Caution is, as always, in order: the Cornell biologist Rick
Harrison has pointed out that when we examine the processes involved in speciation,
we have to be very careful to discern whether we have found the cause *of* speciation or
simply what is caused *by* speciation, whether pattern or a process.

ENDLESS FORMS MOST BEAUTIFUL

We know that speciation occurs, or life could never have diversified in the way it has.
Something that we call speciation indisputably lies at the heart of the enormous variety of living things we see around us on Earth today. Yet, although a huge amount of
energy has been expended on theoretical considerations of the nature of species, and
on the practical business of recognizing species in the natural world, we don't know
nearly as much as we would like about speciation itself. For, even using intensive breeding, short of genetic engineering nobody has been able to produce new reproductive
entities in the laboratory—even though weird and wonderful morphological variations
are pretty easy to obtain by selective breeding. So, while in the lab we can easily mimic
the forces of natural selection as envisaged by Darwin, in a very real sense speciation
remains the "black box" of evolutionary biology. Perhaps this is in part because we
tend to seek a unitary "process" of speciation, when in fact new species are more likely

simply a *result* that we observe in hindsight—a result that may come about through the operation of a whole variety of different factors, ranging from shifts in gene timing and expression, through mutations in structural genes, to chromosomal rearrangements, and even to random behavioral changes.

This would certainly be the conclusion you'd arrive at from reading the huge variety of causes of speciation advanced in the literature for specific cases of reproductive disruption among close relatives. Whatever they may be in a particular case, though, the changes at issue evidently do not necessarily involve the acquisition of anatomical novelties. Or at least, not of the sort that catch the human eye. There is, in fact, a trend among some evolutionary biologists nowadays toward admitting that it's what the animals themselves think (or at least, what they do) that counts. The "recognition" theory of species puts the emphasis on the reproductive signals they send out to each other, rather than on the things that make sense to human taxonomists. It's for them to know and for us to find out—on their terms.

The most relevant question here, though, is less what evolutionary novelties are necessary to disrupt reproductive continuity—for spontaneous genetic changes are occurring in individuals all the time—than how such innovations become transferred from the individual to the level of the whole population. Perhaps this is the moment to mention a famous paradox in speciation studies that was pointed out by the mid-twentieth century biologist and mathematician Joseph Woodger. Paradoxes generally cause your nose to wrinkle a little, warning you that something is not right, and Woodger thought this one incredibly pungent. It is a well known principle in taxonomy that no individual can belong to more than one species. If an individual belongs to species X, it cannot also be part of species Y. Woodger pointed out that when a speciation event occurs, there is logically a point between generations, *between parent and offspring,* where this transition has to happen. If true, this means that if we could take a look at the very brief slice in time where speciation happens, a mother would be in one species and her daughters in another. Not only is this unimaginable, if it did happen, it would raise the classic question of the "hopeful monster" hypothesized by the geneticist Richard Goldschmidt: who would the poor monster mate with?

Fortunately, this is where the inherent "messiness" of nature comes to our rescue. Things don't have to happen as neatly as our reductionist human minds would wish. In his book *Sudden Origins* our colleague Jeffrey Schwartz suggests that most spontaneous mutations in the genetic code occur initially in the "recessive" state, which is to say that they are only expressed in the offspring if they are passed along by both parents. This is highly unlikely to happen early on, when such new alleles will be very rare. If they are thus not expressed, they will not be selected against, and there will be a chance of their spreading "silently" in the population gene pool. Once this covert activity has been going on for a while, a new allele will become common enough to be expressed on multiple occasions. And if it turns out to be advantageous, natural selection can then intervene to favor its spread in the population. Next, imagine this happening in a small population that is isolated from others of its kind for many generations, perhaps thousands of years, and that the genetic change concerned in some way involves reproductive incompatibility with the parent species over the mountain range. Once climate change intervenes to allow the two populations to expand and reunite, the former conspecifics will not reintegrate but will instead find themselves in competition, behaving as separate species. This is, of course, just one scenario. But it does show that geographical isolation can produce reproductive isolation as well as anatomical change and that populations can shift from microevolutionary to macroevolutionary mode vis-à-vis each other. And once more, the disruption of reproductive continuity turns out to be the key.

SO WHAT ABOUT US?

As we hope we have demonstrated by this point, species are in effect the largest interbreeding group, defined among living forms by the ability of members to reproduce with each other. We hope that we have also convinced you that it is reasonable to use diagnosis as a phylogenetic proxy for identifying species in nature. Fortunately we are able to apply both criteria to ourselves, and by both of them *Homo sapiens* is unambiguously a single species: fertile members of one sex of any human population are fully able to reproduce with members of the opposite sex of any other, and no cross-population impediments to breeding amongst humans are known. Individual cases of infertility aside, all humans, from wherever in the world, are reproductively compatible, capable of producing offspring that, given the opportunity, will flourish within whatever social, cultural, or linguistic context is provided by their parents. The vast increase in breeding among people from diverse parts of the planet that has occurred in the past century as a result of global travel and industrialization makes this ever more evident. Unlike lemurs in Madagascar and flies in the Hawaiian rain forest, there is no need to be a peeping Tom to see the biological species concept at work in our species. What is more, there are no fixed diagnostic differences among large aggregates of humans on this planet, either in outward appearance or in genome content, that suggest anything other than a single human species.

Yet, although it is very distinctive anatomically compared to all other life forms on the planet today, our single species is visibly diversified. It is equally evident that this diversity is a product of our biological history. However, in this respect at least, that history is actually both unremarkable and straightforward. Using both the top-down and the bottom-up approaches to determine how species arise and what a species is in nature, evolutionary biologists have determined to everyone's satisfaction that, below the line of death, there are two basic evolutionary processes to which any species is subject.

One of these processes is *diversification,* a clear consequence of Darwin's principle of divergence. In any widespread and successful species—especially one that exists in small, scattered local populations, as *Homo sapiens* almost certainly did until very recently—regional variants tend to develop in local isolates. This is because they accumulate new genetic variations that are not transmitted, or are transmitted with greatly reduced frequency, to their neighbor populations. But environmental and geographical conditions are constantly fluctuating, with the result that a population that is isolated at one moment may be conjoined with its neighbor the next. If and when that happens, any genetic novelties that the isolates might have acquired will become merged into the combined population via the alternative within-species evolutionary process, *reintegration*. This will inevitably take place—unless, that is, speciation has occurred, to isolate the population genetically from its neighbors. This has manifestly not occurred among modern humans, hence the reintegration that has taken place under recent conditions.

Even when we think we might see instances where reintegration may actually result in isolation, we need to be extremely careful in making broad statements about speciation, or even about slight differentiation. A case in point is the interesting situation produced by species that form "rings" around geographic barriers. There are many excellent examples of ring species in modern evolutionary studies. For example, some groups of organisms in Madagascar form something like ring species due to the fact that the island is split by a central mountain range that is currently denuded of vegetation. Most species live in the dwindling forests that ring the outer parts of the island like a green halo. A classic example of ring species can be found in central California, where a central mountain range disrupts the habitat of salamanders in the *Ensatina* complex. David Wake at the University of California, Berkeley, has studied these lizards for decades and has determined that there is a single species that rings the central mountain range. This species has diverged in a sequential fashion from the west, to the north, then to the south, and it has come back into contact in the west.

What happens in such cases is that the ring begins at one end of the barrier and spreads in one direction along the barrier's outer rim. As the species spreads its populations diverge a bit but still retain reproductive contact with each other. Further episodes of spreading and divergence eventually result in a situation where adjacent populations can interbreed with each other, but the two populations at the far ends of the completed ring can't. Voilà! Contact between the two populations that complete the ring reveals reproductive isolation. If we were extremely myopic, we would only see the two populations that complete the ring and would conclude that a speciation event had occurred. However, the daisy chain of population interactions where A can mate with B which can mate with C which can mate with D which can mate with E, but E cannot mate with A, shows otherwise.

On the other end of the spectrum are cases where there is a strong disconnection between what we can see anatomically and what we determine behaviorally. This problem can go both ways. That is, we can find cases where individuals that look very different belong to the same species and where individuals that look very much alike are placed into different species. This is particularly vexatious for paleontologists, who of course only have morphology to go on. That a species may diversify greatly without speciating, or that speciation may take place in a physically undifferentiated species, can make sorting fossil bones into species a tricky task. And the problem traces back to the fact that speciation involves mechanisms which, as we've seen, are often independent of the processes governing morphological differentiation.

In the modern world, the exuberant diversity that may occur within species is beautifully illustrated by the famous photograph showing Christina Aguilera surrounded by Shaquille O'Neal and Yao Ming. It is a lovely photo, easy to conjure up in your mind's eye (if you want to see it simply Google "Shaq Yao Christina"), and it's emblematic of the wonderful diversity *Homo sapiens* displays as a species. An example of the other extreme, where there is little anatomical difference between distinct species, can be found in the famous fruit fly, *Drosophila*. In high school or college many readers will have grown these little guys in bottles in genetics class and will have counted the number of red-eyed and white-eyed progeny in crosses among varieties of the classic species *D. melanogaster*. This species has three very close relatives, called *D. simulans*, *D. mauritiana*, and *D. sechelia*. They are reproductively isolated, yet the outward anatomy is almost identical in all four. The only way to tell them apart is to look at the male penises. Their "whackers," as entomologists inelegantly call them, are shaped differently and are used as the means of diagnosing the species identity of individuals belonging to all four.

"WITHOUT ANY FURTHER INFORMATION . . ."

The twin processes of diversification and reintegration help us see how today's human populations lay nowhere near that region of Hennig's figure 6 where speciation, or complete differentiation, occurs. To demonstrate how important a mechanism reintegration really is, we refer to a passage in Darwin's *The Origin of Species*: "the Negro and the European, are so distinct that, if specimens had been brought to a naturalist *without any further information* [emphasis added], they would undoubtedly have been considered as good and true species." While at first glance this statement might seem to imply that different races or even different species of humans existed on this planet when Darwin wrote, and even perhaps today, the real key to this passage is the qualification we highlighted. The "further information" to which the monogenist and abolitionist Darwin referred, and which he well knew existed and would demolish the hypothesis of separate species, is twofold. First, we know that all humans can interbreed with each other (and do, quite happily and readily). This alone is demonstration of single species status. But, just to ice the cake, we know that *any* aggregate of humans that you might put forward to test the hypothesis of multiple species is undiagnosable as a result of reintegration and hence is unscientific with respect to the way taxonomy and species discovery are approached in modern biology.

To summarize, then, *Homo sapiens* is as well-established a species as exists on the face of the Earth. But although as we'll see in the next chapter its history is not very long, it has clearly been quite complex. For there are clear morphological and molecular signs that, over the past two hundred thousand years or so, both diversification and reintegration have occurred in the human lineage, sometimes concurrently. For most of its history, members of *Homo sapiens* led a hunting-and-gathering life, thinly scattered over vast territories. Such population continuity as there may have been was routinely disrupted by the frequent climatic disruptions of the last couple of Ice Ages.

Small population size, plus frequent isolation, equals perfect conditions for local differentiation. And, of course, for recoalescence as well, when conditions changed—as they frequently did. On balance, however, there was clearly enough isolation for regionally distinctive populations to have emerged over a rather short period of time—perhaps as little as fifty thousand years, as we will see in the next chapter. But at the end of the last Ice Age, things changed. Human beings became sedentary, and coalescence and the associated genetic, linguistic, and cultural interchange among adjacent groups became the norm as populations blossomed. As reintegration became the order of the day, distinctions in all of these domains became blurred, ultimately resulting in the glorious and variegated tapestry of humanity so familiar in the modern world.

EVOLUTION ABOVE THE LINE OF DEATH

Both authors of this book ride the New York City subway every day. During the morning rush hour, a quick glance around any of the cars on the tracks reveals an immense amount of diversity (not counting the rats, pigeons, and other interesting wildlife one is prone to encounter on the subway). We all look very different from each other and it isn't because we haven't had our coffee yet. It's because all of the *Homo sapiens* on the subway *are* different. Since it is the differences amongst people that we are most likely to use to categorize and classify people, it is extremely important that we delve into how humans became so different-looking. How did we get this way? The simple answer is we evolved. And so at this point, we need to pause to consider the process of evolution itself. And one of the best ways to describe how evolution proceeds is to show how it *doesn't*. Because we are very interested in our own biology and how we got this way, we are very prone to focusing excessively on ourselves. As the only organisms on the planet that can think about our origins and our biology we are too often excessively in awe of our own specialness. As a result of culture, or religion, or our social ideas, we are tempted to think that we are optimally designed. After all, we are the only species on the planet that can think about thinking—surely the result of some major fine-tuning. However, consider the possibility that we are simply a mistake or at least the outcome of a chance event. Or, maybe worse, perhaps our species *Homo sapiens* actually arose and evolved as the result of a whole chain of mishaps or chance events. Not the most ego-gratifying of conclusions, perhaps, but as evolutionary biologists we must confess to preferring this scenario, not because we are Debbie Downers, but because of the very ways in which evolution works.

To put it in a nutshell, it is evident for a whole host of reasons (ranging from the structure of our knees to our decision-making processes) that we humans are not optimally designed, and the underlying reason for this is that we are just one more species in a great tree of life that has diverged and branched from a single common ancestor that lived over 3.5 billion years ago. We weren't designed; we evolved. And our brains, muscles, bones, physiology, behaviors, and many other of our characteristics bear the scars (and sexy curves) of this long, multistage process. Of course, at this point it might seem natural to ask, "Didn't natural selection craft the look and feel of all organisms during this grand tree-of-life experiment? And if so, doesn't natural selection select only the best solutions to the evolutionary challenge?" Well, actually no. Or at least, more no than yes. The matter turns out to be pretty complex.

WHY WE HAVE MESSY BODIES

Let's start at the beginning, which means looking at natural selection itself. To recap, natural selection is the process proposed by Darwin whereby the "fitter" reproduce more successfully than their less well-adapted fellows. However, while it may push traits in specific directions, natural selection does not push them toward perfection, and there are at least four reasons why it doesn't.

First, the "designs" of organisms do not arise *de novo*, as human designs for computers or libraries can do. Both of us were at the American Museum of Natural History some twenty years ago, when its library, which has one of the most amazing collections of natural history volumes in the world, still used an archaic, incomprehensible but quaint numbering system for its books. This numbering system was a local hybrid of the Universal Decimal Classification (UDC) system and had been in place since the founding of the museum more than one hundred years earlier. Over the years the librarians had modified it to fit their needs. Anyone not familiar with the resulting arrangement (that is, almost everybody except the librarians and the most emeritus of curators) would find themselves fruitlessly wandering around the stacks for hours looking for books and periodicals. In 1995, the AMNH librarians finally removed the old system entirely and adopted the Library of Congress system. Because the librarians did not want to be encumbered with any of the characteristics of the old numbering system, they simply got rid of it wholesale and imposed the new, entirely unrelated system. Bingo! At last it was possible for the average person to find books easily in the AMNH library. This was clearly the optimal solution.

But evolution does not—and cannot—work this way. The hereditary information of each organism and species is carried in its genome, and because massive resetting of genomes during evolution simply does not happen because of the way these intricate entities work, a strong historical contingency is necessarily carried on from generation to generation, even as evolution continues on its merry way. Natural selection thus can only work with the forms it has at hand; it cannot request a "do-over," or what golfers term a "mulligan"—although, as we will later see, the way genomes work often "unwittingly" allows natural selection a second chance at making things better.

A second reason why natural selection does not necessarily fashion perfect adaptations is that not all changes in evolution are implemented by natural selection. Some changes are just the result of plain good (or bad) luck that happens completely by chance. Natural selection's twin, if you will, is a process called random drift. Drift happens in small populations because of a phenomenon called sampling error. You can think of this in the following way. Take four hundred coins and flip them. While it is possible, you would be rather foolish to bet that you could flip four hundred heads in a row (the probability that you would get four hundred heads in a row is ½ raised to the 400th power; that's a pretty big number). The sober reality of four hundred coin flips is that you would see a lot of heads and a lot of tails (on average two hundred of each). Now take four coins. In this case it would be a relatively good bet that you could flip four heads in a row and not see any tails (the probability of getting four heads in row is ½ raised to the fourth power, so you would be successful one in sixteen times you tried). This identical sampling error principle also works for sperm and eggs in natural populations of breeding organisms. And it works no matter what kind of organism we are talking about.

Here is one of our favorite examples of why we need to be careful about making claims about adaptation and natural selection. It concerns the strange case called "guevodoces," which crassly translates as "balls at twelve." This genetic syndrome concerns a metabolic change in genetic males who have a deficiency of an enzyme called 5-alpha-sterol-reductase. This enzyme is responsible for the accumulation in the male body of the hormone testosterone, from its precursor molecule dihydrotestosterone. Male fetuses with the syndrome do not have enough testosterone in their endocrine system to trigger the development of a penis from their genital tissues. As a result, these males are born with what look like external female genitalia—they are born as little girls. This syndrome actually existed at one time at a surprisingly high frequency in isolated villages in the Dominican Republic. And it turned out that all of the males with this syndrome could be traced back to the same great-great-great-great-great grandmother, a woman named Altagracia Carrasco. Naturally enough, these individuals were treated as little girls from their birth onward. But a strange thing happened when they entered puberty. By that point enough testosterone accumulates in the bodies of individuals with this condition to circumvent their 5 alpha sterol reductase deficiency. They then dramatically and rapidly grow a penis, and their testes descend—hence the vernacular name for the condition. These males are fully fertile, virile, and often go on to reproduce. For the most part, the unusual developmental progression of these individuals was not looked on as strange in any way at all by the villagers; after all, they were used to it. But it seems odd to us, to say the least. So why did this syndrome arise in this population, and why has it persisted?

Well, if we wanted to, we could make up a pretty good story here about natural selection involving a stride toward perfection. We could be sociobiological about it and say this syndrome persisted because these males are better at relating to women, having spent the first part of their lives as females. Because they can relate to females better, isn't it reasonable to assume that they might make better mates? Right, ladies? Because they were preferred by females, they would be able to reproduce more than other males, and natural selection would drive the condition onward. An attractive idea, but alas a wrong one. The real reason for the origin and persistence of this syndrome is basically a random one, relating to the fact that all of the individuals involved are descendants of that one individual, Altagracia Carrasco. And since they were not at a reproductive disadvantage, their genotype was not winnowed out of the population.

Every genetic novelty arises as a result of random chance (copying error), and sometimes strange things hang around in populations simply because they don't get in the way, as in this case. Of course, we have set up a straw man here, but there are many instances of serious scientists making not-dissimilar claims about natural selection, without really digging deeper into the biological reasons for the existence of particular traits in populations.

What drift means with respect to our adaptation is that much of what we observe about ourselves is there by chance. And what this has meant over the longer term is that, as we evolved, a lot of the changes that accumulated were the result of chance events, built on top of more chance events, with some natural selection tossed in. Not a very efficient way to produce a species, or any product for that matter. But the key is that it works. At every stage the organism is fully functional. And, as we will see, it is probably why we have the remarkable brain we have, rather than a soulless computer between our ears. It's also why we walk upright, but with incredibly bad knees and ankles, and it explains why most of us live in groups, but strive to retain our individuality.

A third reason why natural selection does not result in perfect organisms or structures lies in the workings of the hereditary mechanism itself, in which every gene is involved in multiple functions. So while some adaptations appear to be pretty good if not perfect responses to particular environmental challenges, having them may produce adverse effects on performance in other traits. Take, for example, the strange case of a large Hawaiian fly called *Drosophila heteroneura*. Its very name implies that something is going on with its head, and indeed the male flies in this species have heads shaped like the business ends of hammers, in which the eyes are on the end of thick stalks. Why? It turns out that the more hammerheaded a male fly is, the more successful he will be at reproducing. Why again? Two reasons. The first is that female flies seem to prefer males with broader heads. In other words, these females find males with large hammerheads "sexier" than males with smaller, narrower heads. The second reason is that males with bigger hammerheads can actually fight better over the available females, hammering the opposition as it were. This gives the big-headed males a double sexual whammy. On the other hand, these male flies with the big broad heads are also most likely at a disadvantage in daily life, because they become very conspicuous, thus making them more susceptible to predation. It is also possible the broad head could actually impair their vision or movement and thus preclude the flies' escape from predation. If you are eaten, it doesn't matter how sexy you are to a female of your own species.

The final reason natural selection doesn't make perfect organisms is that, as we've already noted, natural selection works on existing variation. The genomes of organisms cannot just "conjure up" new variants that might be good at responding to environmental challenges. However well some organisms and viruses may appear to do at accommodating to such challenges, this apparent adaptation is not the result of consciously creating new variation and using it. For example, HIV has one of the most rapid rates of response to antiviral challenge shown by any virus. This is not because the HIV virus can sense the kind of changes it needs to make in its genome. Rather, as an RNA virus, it naturally mutates rapidly. This rapid mutation randomly generates the genetic variability that is then acted upon by natural selection (in this case, antiviral medications). Many bacteria also adapt very rapidly and efficiently to environmental challenges such as those posed by antibiotics. Again, this is not because the bacteria "sense" that they need variation that could help in developing drug resistance, but rather because among bacteria variation is prevalent in the form of plasmids that can carry drug resistance.

In a rather similar way, we humans also have something about us that is unique in the world of natural history as far as conjuring up variability is concerned. This is human culture, which introduces the possibility of cultural evolution. But in contrast to the examples just given, cultural variability is easily "conjured" up—as evidenced, among other things, by changing clothing styles and musical tastes. As we will see later, cultural aspects of our evolution have greatly affected how we interact with the environment and how our psychology works, sometimes outstripping the effects of Darwinian evolution.

STUCK IN A GENETIC MESS?

While viruses and bacteria are famous for their ability to adapt quite rapidly and efficiently to challenges from antibacterials and antivirals, most evolution on this planet proceeds quite differently. Our genomes, and the genes that code for the traits that make us who we are, are transmitted in such a way that huge changes that could potentially be directed to "perfecting" ourselves are nearly impossible. By perfection we simply mean the use of the genetic information to produce the most efficient system possible.

Let's look at our brains as an example. It has been estimated that over half of the genes in our genomes are involved in brain structure, function, and development. This means that some twelve thousand genes are involved in a very intricate dance that, if disrupted, will produce a pretty messed-up creature—even more messed-up than we are normally. In addition, because humans nowadays live in large populations, we wouldn't expect random drift to exert itself very much if at all: it is most effective in small populations. And genetic drift might actually be the best mechanism to explain large changes in evolving brains. Indeed, small populations and chance events in our past are most likely the reason our brains are the way they are now. For our brains are not finely burnished decision-making machines. The cognitive scientist Gary Marcus has recently published an entertaining book entitled *Kluge* about all of the ways in which we humans contrive to make bad decisions and continue to believe weird things in the face of all contrary evidence, and no human being reading it can fail to resonate to many of his examples.

Does any of this mean that human brains, and in the larger context humans themselves, have stopped evolving since we became *Homo sapiens?* Does it imply that the human brain and our bodies will remain untidy for ever? These questions are complex, and the answers are conditional upon many things. Slight changes in our responses to the intense modern-day stimuli our species experiences might (or might not) have occurred over the last few decades. But large-scale changes—that is, quantum jumps in our brain structure and function and in our overall anatomy, physiology, and behavior—are probably not going to occur in our species as long as we remain in huge, densely packed populations with highly mobile individuals. The basic mechanisms of genetics simply won't allow such large-scale changes to happen. Even if an extremely advantageous mutation were to arise in a human population that gave its owner a huge advantage, such a mutation would more than likely sooner or later be swamped by all of the normal alleles in the population. This is both the scourge and the beauty of being a mobile species with large populations. In concert with natural selection that keeps it within limits, our variation ensures that our populations float around the mean for traits. This isn't bad news, because this is the process that keeps our species healthy. For, as highly advantageous genes go, so also go the genes that make us sick.

Part of the story of how our brains and our existence got so untidy is written in the history of life on this planet over the past 3.5 billion years. This history is stored in the genomic information of every organism. From the smallest, most insignificant bacteria or archaeon, to the larger and seemingly more complex organisms on the planet such as ourselves, there is an unbroken chain of descent from a common ancestor. And by knowing the who, the what, the when, the where, and the how of these relatives scattered throughout the tree of life, we can trace how our brains and the brains of other organisms evolved, following the rules we've just explained. By grasping the history of how brains evolved, and how our ancestors (not just the ancestors in our immediate family, but our ancestors all the way back to bacteria) functioned, we will have a better chance of understanding how our brains work.

The problem of reconstructing evolutionary histories is rather like a Rubik's cube: solving it takes a large number of moves. And while there is only one solution (all sides end up the same color), there are many ways to arrive at it. There is a best way, and a quickest way, to solve the puzzle, and this would be the perfect solution. But even if, as likely, you solved it after many more moves than the minimum, you made only one set of moves. And if we wanted someone else to solve it exactly the way you did, that person would need clues to the moves you made. In the evolutionary puzzle, it is the ancestors in the tree of life that provide the clues to what happened. Almost certainly, the solution that these ancestral clues lead you to won't have been the theoretically perfect one, but with luck it will be close to the convoluted one that the evolutionary process actually took.

THE NAME GAME

Humans have named objects in nature since the beginning of language. In this respect, the process or science of taxonomy must have ultimately begun with language. Folk taxonomy has been a favorite subject of anthropologists and ethnologists since the 1960s, and the analysis of names given to things in the natural world exerts a perennial fascination. Carol Yoon suggests that the propensity of humans to name things is somehow "hard-wired" in our brains. It is there, Yoon suggests, to cope with what she calls the "umwelt," or the realization that we are part of the natural world. Because, as we saw in the opening paragraphs of this chapter, there is a lot of apparent chaos all about us with respect to the umwelt, the naming of things becomes an important tool for organizing and interpreting the beautiful mess by which we are surrounded.

One of the first recorded thinkers about the naming of things was Aristotle. That many of his writings have survived to this day testifies to the strength of his approach. Aristotle's vast writings addressed many aspects of animals, and it is obvious from reading his *Historia Animalium* and other writings that his grasp of organismal diversity was quite extensive. In the *History of Animals* Aristotle described the anatomy, modes of existence, and activities of over five hundred animals familiar to him and other ancients residing in Greece. Aristotle's way of naming things was very utilitarian. When he spoke of a *uev* (bee) he had to be sure that his readers were aware that he was referring to a bee and not a spider or, worse yet, the close relative of bees, *aKpis* (wasp).

There are up to four hundred species of primates. They all have names (or at least the great majority do). Most of the names are strange-sounding, and to the untrained ear most of the names will not make sense. Yet there is a great deal of order to the names that have been given to these species. Names are given to species and other natural units according to a set of well-recognized rules. Scientific names may reflect the appearance of the thing, or signify the geographic area where the thing is found, or honor a fellow scientist, or be given for a myriad other reasons. But while it's an ancient process, the giving of names to things has only recently been unified in science. So what might at first seem like a simple procedure may actually be a very difficult and involved process, with both historical and philosophical underpinnings.

As we saw in the last chapter, the most significant advance in naming things came from Linnaeus, the "father of taxonomy." Linnaeus excelled at logic in school and was a renowned numerologist. Ernst Mayr suggests that both of these aspects of Linnaeus's intellect were important in his contribution to zoology. Linnaeus, who was often described as arrogant and self-promoting, is sometimes also chided for his inflexibility and pedantry, two accurate descriptors of his intellect. In some ways, his emphasis on nomenclature and classification was a handicap to any larger view, but his pedantry and rigor were essential to his system of nomenclature. As we've seen, he established the binominal system of names that we are all so familiar with and further established a logical and mainly consistent way of naming and classifying things. The rigor of Linnaeus led directly to the establishment of the codes of nomenclature that all taxonomists—botanical, zoological, and microbial—still abide by.

So far we have used the words taxonomy, systematics, and classification quite loosely. What, then, are the differences between taxonomy and systematics, between taxonomy and phylogenetics, and between systematics and nomenclature? Several authors have attempted to discriminate between and among these terms. G. G. Simpson and Ernst Mayr, two of the formulators of the New Synthesis, discriminated between taxonomy and systematics by suggesting that taxonomy is a subdiscipline of systematics. Mayr gives strong and clear definitions of these terms (see text box, p. 85). His definitions imply that taxonomy is a subdiscipline of systematics, that the subjects of classification are organisms, and that the subjects of taxonomy are classifications. We might note in passing that the embedding of taxonomy within systematics has been objected to by some authors, notably those who turn these definitions on their head and suggest that taxonomy is the broader in scope and that systematics is a subset of taxonomy. In this alternative definition, taxonomy not only includes the naming of organisms but also subsumes the deciphering of their relationships to one another.

While the three to four hundred species names that one sees in the mammal order Primates may seem pretty random, their names arose from one of the most rigorous

Mayrian definitions of taxonomy, classification, nomenclature, and
systematics:

Systematics is the scientific study of the kinds and diversity
of organisms and of any and all relationships among them (p. 7;
from Simpson, 1961).

Zoological Classification is the ordering of animals into
groups (or sets) on the basis of their relationships, that is, of
associations by contiguity, similarity, or both (p. 9).

Zoological Nomenclature is the application of distinctive
names to each of the groups recognized in any given zoological
classification (p. 9).

Taxonomy is the theoretical study of classification, including
its bases, principles, procedures, and rules (p. 11).

and self-policing procedures in all of science. Rules for the naming of animals exist in
the form of the International Code of Zoological Nomenclature (ICZN). The ICZN
reads like a constitution, with articles and amendments, and is used to maintain some
semblance of order, continuity, and consistency in the naming of animals. As the science of taxonomy progressed, and the number of taxonomic ranks grew, the need
became clear for standards to exemplify the taxonomies proposed. Each larger group
needs to be exemplified by a smaller one, starting at the bottom with an individual. For
instance, each family has a type genus. For our own family (Hominidae) the type genus
is *Homo*. Each genus has a type species, and in this case, unsurprisingly, *Homo sapiens* is
the species tagged for this job. Every species must in its turn have a type specimen, and
this will be a specified individual, or perhaps group of individuals from the same place,
most likely residing in a museum collection. This individual will serve as the definitive
example of the morphologies that characterize the species.

So, you must be asking, what—or more appropriately, who—is the type specimen for *Homo sapiens*? Well, appropriately enough the type specimen's name is, in its
Latin form, Carolus Linnaeus, and the type locality (which refers to the geographic
location where the type specimen was obtained) is Uppsala, Sweden (Carl Linné's
place of residence and burial). The holotype of *Homo sapiens* therefore resides in Carl
Linne's coffin in the graveyard of the Uppsala Domkyrka (Cathedral of Uppsala).
Despite recent attempts by some misguided souls (which actually date back to the

proposed type specimen's own efforts in the late nineteenth century) to designate the paleontologist Edward Drinker Cope's skeleton as the type specimen of *Homo sapiens*, Linnaeus's corpse holds the title as a result of the law of priority. Remember that Linnaeus named the species *Homo sapiens* in his *Systema Naturae*. He did not designate a type specimen because this was not normally done by natural historians at that time. However, in 1959, on the occasion of the two hundredth anniversary of the publication of the tenth edition of *Systema Naturae*, W. T. Stearns proposed Carolus Linneaus as the type specimen of our species. Stearns justified this choice by noting, "Since for nomenclatorial purposes the specimen most carefully studied and recorded by the author is to be accepted as the type, clearly Linnaeus himself, who was much addicted to autobiography, must stand as the type of his *Homo sapiens*." The logic is clear, although Stearns does no more than imply that Linneaus's arrogance and self-absorption would have made him the specimen of *Homo sapiens* he had studied the most carefully—and thus would by default be the type specimen of our species. Not that this makes the type terribly useful, for you would need to exhume poor Linné every time you wanted to make a comparison. Cope, on the other hand, had the foresight to have his corpse (posthumously, of course) stripped of its flesh and stored in the University of Pennsylvania's Wistar Institute for future researchers to study. A lot of effort on Cope's part for nothing, but so goes the law of priority in taxonomy.

One of the major problems that the modern biological world faces is what has been called the "taxonomic imperative." This problem exists because the attempt to name and describe all of the ten million–plus species that are living on the planet is failing. It is failing not because a good system for naming things doesn't exist, nor because scientists aren't trying, but rather because there just aren't enough taxonomists on the planet to get the job done. There are two obvious possible solutions to the problem, and both should probably be applied. We can train more taxonomists, or we can speed up the process of taxonomy.

In a world of shrinking resources for science, the latter solution has received a lot of attention in the last decade, and the internet and world wide web have also greatly enhanced the speed with which things are getting named. One approach in particular, using DNA sequences, has been suggested as the panacea for solving the taxonomic imperative. It has been called "DNA taxonomy" or the "DNA barcode of life" by its proponents and the "barcode of lies" by its detractors, who are mostly the taxonomists working in the trenches using more traditional techniques. It is very illuminating to take a close look at this initiative, mainly because it has faced the very same problems that we have been discussing in this book with respect to taxonomy, species delimitation, and species patterns and process. So let's examine how molecules are used in systematizing and naming organisms. Our excursion into the realm of DNA barcoding might seem a bit off the mark at first, but it really is at the heart of how we use molecular and genetic information in understanding the diversity of life on this planet of which humans are just one species.

USING GENES TO NAME THINGS

The use of molecular markers in studies of evolution has been a tried and true approach for some time now. It started as far back as the early twentieth century, when blood types were used to examine the population genetics of humans, and it saw a rebirth in the 1960s, when Richard Lewontin, Jack Hubby, and Harry Harris applied the approach of electrophoresis to the proteins of flies and humans. This procedure required that sera or tissue be taken from individuals and ground up. In the case of the poor fruit flies, the entire fly was ground up. By grinding up the fly or the tissue, proteins were released from the cells of the subject. Proteins are made of amino acids that have electrical charges on them. When these proteins are placed in a slab of a flat gel (a starchy substance usually isolated from potatoes) and an electrical current is applied, the proteins will move through the gel. How far they migrate is proportional to their electrical charge (which sometimes also correlates to their size). If the same proteins taken from two individuals are identical in their amino acids, then both will migrate the same distance in a gel. If a mutation has occurred, and the amino acids in the protein are slightly different between two individuals, the proteins will usually migrate different distances. In this way, differences in proteins may be detected between individuals within populations and indeed between individuals from different species. Such molecular differences turn out to be much like the anatomical characters that traditional taxonomists work with.

Nonetheless, some tension arose in taxonomic circles as a result of this new technique. Taxonomists called the scientists who used electrophoresis "gel jockeys" or, worse, "find 'em and grind 'em" scientists. Taxonomists viewed their science seriously, and they felt that the electrophoresis results had little if anything to do with their trade. The problem was, of course, one of unfamiliarity: after all, for about 150 years taxonomists had been doing things quite well without molecules. Soon, however, most of the evolutionary field started to use electrophoresis, and traditional taxonomists were seen as Luddites when they failed to incorporate the new technology into their trade.

The 1980s saw the infusion of even newer technology into evolutionary biology. DNA sequencing techniques were developed to examine population and species level phenomena in evolutionary biology. The UC Berkeley lab of the late Allan Wilson was the early Mecca for applying molecular techniques to the study of evolution. Always on the cutting edge because of the sheer number of people there who had keen interests in phylogeny, population biology, and evolution, Wilson's lab always seemed to be a few years ahead of everyone. Indeed, some of the first papers to report on the use of the new generation of molecular tools in evolutionary biology, namely DNA sequences, also came out of Wilson's lab.

The main reason for the proliferation of DNA sequence data in systematics was the polymerase chain reaction or PCR, a method invented by Cary Mullis, who had worked with Wilson and others at Berkeley. In fact, one of us (Rob) witnessed the early infancy of PCR in Wilson's lab without even recognizing the significance of the approach. For most of December 1984 to July of 1985, three water baths set at different temperatures, and sitting on Russ Higuchi's bench, were the focus of much mystery. Higuchi was one of Wilson's postdoctoral associates and had done the hard work of cloning and sequencing a gene from the extinct quagga. Higuchi had several undergrad interns who would sit in front of the water baths with a stopwatch and move tubes from one bath to the next every minute or so. When the third bath was reached, the student would move the tubes back to the first bath, then to the second, and then the third, and this would go on for about two hours, when the student would put the tubes in a freezer and leave. When asked what he was doing, Higuchi would just wink and say "an experiment for Allan." It turns out that these undergrad interns were actually the prototypes for the PCR machines, or thermal cyclers, that do the job of PCR today: machines that would eventually reside in the lab of nearly every biologist interested in looking at genes.

Other novel approaches were developed in the 1990s that immensely sped up the process of DNA sequencing. To illustrate the extraordinary rate at which the technology has advanced, we point to one exception to the Wilson lab's stranglehold on molecular technology. It comes from a paper that was published by the University of Chicago biologist Marty Kreitman, then at Harvard. Kreitman sequenced a gene responsible for clearing the body of alcohol in eight fruit fly individuals. It took him four years and a lot of effort to accomplish the work for this important paper, but while this publication was a decade ahead of its time, only two decades after its appearance a student wishing to redo this research could accomplish it in one day. Average sequencing power has increased about at the same rate as Moore's law for computers, which states that the average computing power will double every eighteen months. Now, with next generation sequencing recently impacting DNA studies, the entire genomes of several humans have been obtained. We will shortly discuss the ramifications of this new technology on our understanding of race and disease.

MOLECULES AND SPECIES

Using DNA sequences to identify organisms has actually been around for some time. Carl Woese, a University of Illinois biologist who has the great distinction of discovering an entire new super-domain of organisms (called Archaea), was probably the first to see the merit of using DNA sequences to identify microbes. And in forensics, Alec Jeffries pioneered the use of DNA variation as a means to identify human individuals. Other animal-oriented forensic applications were also pioneered in the 1990s, such as using DNA sequences to identify the species that produced different kinds of caviar and studies on the species origin of meat products from marine organisms. In 2003, the approach for nonhuman identification was formalized under the banner of DNA barcoding, when University of Guelph biologist Paul Hebert suggested that a small DNA sequence, from a uniformly chosen gene, would be useful in obtaining identifiers for all of the named species on the planet. Hardcore taxonomists were, needless to say, a bit upset by the intrusion of DNA into their everyday lives. By taking a close look at why they were upset, we can better understand what it means to discover new species and hence understand better what part of the problem is with looking at the genealogy or ancestry of humans.

So how do taxonomists delimit new species? As Diana Lipscomb of George Washington University and her colleagues point out, the process for doing this is in principle a rigidly intellectual (and purely scientific) one, involving proposing hypotheses and subsequently testing them. First, you put forward the notion that a particular group of organisms composes a unitary species. Then you test this notion by determining whether the group is indeed differentiated enough from its closest relatives to describe it as a new species. In the context of scientific endeavor, this is about as pure as any science can be. The big question is, of course, what "differentiated enough" actually means.

DNA barcoding, as executed by the majority of its practitioners, attempts to determine what "differentiated enough" means in a purely inductive way. They take observations on well-known species and estimate how similar they are to other species. In this way, the barcoders suggest, they can come up with what they call a "DNA barcode gap" or a calibrated genetic distance that they then use as a cutoff point for species. In other words, if a barcoder is looking at a group of closely related species, and observes that the nearest species is 2 percent different from the species of interest in the gene sequence observed, then he or she will use the 2 percent difference as the barcode gap for recognizing species-level differentiation. Another approach is to use the evolutionary biologist's trusty method of tree-building, in which forkings at the base of the tree represent taxonomic branching amongst genera and families, whereas the branches closer to the tips of the tree represent species-level differentiation. Tree-like diagrams of this kind can be produced from DNA barcoding information. Then the structure of the tree is evaluated, and an arbitrary line is drawn across the tips of the tree. Anything above the line that collapses to a single point is considered a species. The arbitrary line that is used in this approach is much like the observed distance that established the DNA barcode gap in our earlier example.

Both of these approaches sound reasonable—right? Well, not really. Part of the reason why evolutionary-tree research went through a revolution in the 1960s and 1970s was because of an attempt to eliminate the arbitrariness of the choices made by taxonomists of the day. Those taxonomists were known as "evolutionary taxonomists," which doesn't sound so bad. After all, they were using the notion of evolutionary relatedness to delimit and name species, and their very name reflects a supposed reliance on evolution to make those decisions. But not so fast! The evolutionary taxonomists were actually using their own personal judgments in making these taxonomic decisions. They resorted to what can best be called "expert testimony" to make them. And if you were outside the core group of "experts," then your opinion was unlikely to be heard. Everything depended on the personal authority of the taxonomist.

The cult of authority was bound eventually to produce a reaction, and the ensuing revolt against the evolutionary taxonomists turned their title into one of insult. If you were called an "evolutionary taxonomist" in the late 1970s, this was usually not a polite thing. Evolutionary taxonomy simply wasn't considered science, because it was authoritarian and most of the time was not reproducible. In keeping with the social revolution of the 1960s and 1970s, evolutionary taxonomy yielded to a more "democratic" approach to doing taxonomy and systematics. The new approach was not democratic because lots of scientists got together and voted on the outcome. Rather, it received the epithet because the basic data, and the observations made about nature, were allowed to "vote" on the outcome. This revolution in systematics came about from the development of techniques that made it easy to count such votes (observations) about historical events and to declare a "winner" on the basis of the evidence at hand. If a researcher collected data relevant to a taxonomic question and then analyzed them using these methods, anyone else could take the same data and get the same exact answer. Objectivity was rallying cry of the day, and of course it should still be. But both the DNA barcode gap and the species lines drawn on trees actually hark back to evolutionary taxonomy. And one reason why the classical taxonomists were so vociferously against these new techniques is that many of them had fought in the trenches to establish the objectivity of the current taxonomic methods. To watch their discipline revert to the state that had existed forty years earlier was a bit like an old hippie watching a George Bush take the oath of office, three times.

So how would the classical taxonomists use DNA sequence data? They would use them to help establish diagnostics for the species being considered, just as they had used anatomical characters to find diagnostics for species in the old days. Anyone who has ever used a "key" to figure out what plant or animal they are looking at knows both what a diagnostic is and what makes keys work. Most keys are dichotomous, meaning they give you two choices at each level. So at each level, two descriptions are given. If one of the descriptions is accurate for the taxon you are trying to identify, then you read either the name of the taxon, or you are directed to examine another attribute of the taxon involved. The diagnostics are the attributes you are directed to examine, so a "diagnostic" is an attribute of a taxonomic group that allows for the identification of that group.

The most basic key that we use for the organisms on this planet involves the discrimination of the three super domains of organisms—Eukaryota, Bacteria, and Archaea. With three taxa such as these superdomains, the key can be constructed many ways. We can start with the classical divide between the three domains, whether an organism has cells with nuclei or not. If they do, then they are eukaryotes, and if they don't then a second set of questions is posed. Table 1 is a key for the three domains of life on the planet starting with eukaryotes as the first to be identified.

TABLE 1. Taxonomic key for the three domains of life where prokaryotes are considered a good taxonomic group

1a.	Organisms with cell nuclei	Eukaryota
1b.	Organisms lacking cell nuclei	2
2a.	Cell membrane has glycerol‑ester lipids and reverse stereochemistry of glycerol	Archaea
2b.	Cell membrane has glycerol‑ester lipids and normal glycerol stereochemistry	Bacteria

Another way to start the key is to peel Bacteria off first, with the initial statement in the key. In this case, the first step would be to determine if the genes in the groups lack what are called introns. Introns are stretches of DNA inserted in the coding regions of genes. These interrupt the gene and do not code for protein, like the rest of the gene. Instead, they are processed out of the gene before the gene is translated into protein. It turns out that both Archaea and Eukaryota have introns in their genes, albeit not the same kind of introns. The second statement in the key would then use either the presence of a nucleus in the cell, *or* whether or not the cell membrane has glycerol-ester lipids. That key would look like table 2.

And yes, you guessed it, there is a third way to generate this most basic of all keys: start with Archaea by using the presence of glycerol-ester lipids in the membrane

TABLE 2. Taxonomic key for the three domains of life based on the current best understanding of domain relationships

1a.	Organisms with genes lacking introns	Bacteria
1b.	Organisms lacking cell nuclei	2
2a.	Cell membrane has glycerol‑ester lipids and cell has no nucleus	Archaea
2b.	Cell membrane has glycerol ester lipids and cell has a nucleus	Eukaryota

and then, once you have determined whether you have Archaea or one of the other two, use the presence of introns to discriminate between Bacteria and Eukaryota.

All this goes to show that classification keys can be generated even for "unnatural" groupings of organisms. Unnatural groups consist of organisms that don't have a single common ancestor to the exclusion of all others. The great example of an unnatural group of organisms is Prokaryota, which brings together the Bacteria and Archaea. Grouping these two together is the old way of classifying cellular life that you probably learned in high school. The bottom line here is that all you need for a key to work is a diagnostic—even if the group is not a natural one. So be cautious! The keys get a lot more complex by the time you reach the tips of the tree of life. And the attributes used to key out the organisms become more and more specific the further out you are toward those tips. What's more, the more closely related the taxa are, the more attributes they will generally have in common, and common attributes don't help much in keying out closely related organisms.

There is a simple rule for how a taxonomist finds good diagnostics. Any attribute of a group of organisms that is fixed, and different from its counterparts in other groups of organisms, may be a diagnostic character. Thus if a group of organisms called species A all have a nose with a finger coming off it and no other species shows this attribute, then a nose with a finger coming off it is a diagnostic for species A. Likewise, if we look at the sequence of a gene of ten individuals from species A, and all ten have an A in the first position of the gene whereas in the closely related species B we look at ten individuals and all ten have a G in the first position of the same gene, then the A becomes a diagnostic character for species A, and the G becomes a diagnostic character for species B. This process works well for taxonomists using anatomy and other visible attributes.

One snafu that may arise in using this approach is that if you don't look at enough individuals in a group, you will tend to "overdiagnose" it. Let's say you examine what you think is species A, and you find nine individuals with an A in the first position of a gene, and one individual with a G in the first position of the same gene. Species B has nine individuals with a G in the first position, and one individual with an A in the first position. In this case, that attribute is not a diagnostic character, and you cannot say a diagnostic exists for the two species. Now let's say you have only half of the funds necessary to collect and process enough specimens from species A and species B. Instead of analyzing ten individuals, you can only collect and analyze five. In this case, you will most likely see an A in the first position of the gene for all individuals of species A and a G in the first position of the gene for species B. In this case, you would conclude that the first position of the gene is a diagnostic character for both species, and you would have overdiagnosed the two as distinct if you based your taxonomy only on the first position of the chosen gene. The problem of overdiagnosis also extends to when you don't look at enough groups of things. What's more, it is not solely limited to finding diagnostic characters. The inadequate sampling of organisms will always be a potential problem in any endeavor related to understanding change at the level of species. Whether or not DNA barcoding will in the end settle on using diagnostics, or stick with using distances, is a question that will be answered ultimately by the barcoders. But the controversy between traditional and barcoding methods has clarified the problem of classifying things in nature.

The same diagnostic principle can also be used in a hypothetico-deductive context to ask questions about species discovery. The resulting process is called population aggregation analysis (PAA), and it involves testing successively aggregated populations that are considered to be species. If diagnostics can be found for a particular way of aggregating individuals, then a strong case can be made that the aggregates are species. The key question surrounds the phrase "for a particular way of aggregating individuals." The process of aggregating individuals needs, at its core, to be hypothetico-deductive. We simply can't go around aggregating individuals because we observe some *difference* amongst them and then use that difference as a diagnostic. That would be circular and terrible science. Instead we have to aggregate individuals based on geography or behavior or some other positive attribute that produces a hypothesis to be tested with independent data. That's how science works.

Any hypothesis generated this way must, of course, also make sense with respect to a species concept. After first hearing about the PAA approach at a meeting, a very famous population geneticist boldly said, "This PAA thing simply won't work because I can ultimately create an aggregate that would include me and my offspring. I am certain I could find a DNA sequence marker that would diagnose me and my offspring as distinct from all other humans and therefore my offspring and I would be classified as another species." This statement has much truth to it. After all, it is true that even a single individual human can be identified with a DNA sequence fingerprinting test, as we see every week on television programs such as *CSI*. Forget entire DNA sequences: every single individual human has a unique DNA fingerprint. Still, the famous geneticist's statement is actually relevant if, and only if, the aggregating of that scientist and his offspring is warranted by the testing of the hypothesis that they belong to different species. And of course, no one in his right mind would suggest that the scientist's family belong to different species, not even the scientist's (or his offspring's) staunchest critics. Again, this is where having a species concept (in contrast to species criterion) is critical. Any test of an aggregate that violates the concept being used is an invalid test. In this case, our famous scientist and his offspring are still part of the same species under the biological species concept articulated by Mayr. They could reproduce with each other if they wanted, provided they were of the appropriate sexes. But then again, maybe they wouldn't reproduce, or couldn't because they now live on separate continents. Even more farfetched would be that the offspring *have* become a new species. In that case the original statement would be correct, but then again, this would return us to Woodger's paradox.

WHAT'S THAT ODOR? THE RETURN OF WOODGER'S PARADOX

As we noted earlier, Joseph Woodger suggested that, when a speciation event occurs, there should logically be a point between generations, *between parent and offspring,* where the transition from one species to the next has to happen. This paradox could really pose a problem if Woodger was right, because if we could take a look at that very brief slice in time where all of this happens, we would find that a mother would be in one species and her daughters in another. In order to understand why Woodger's paradox is less odoriferous than it seems at first, we need to delve a little into the dynamics of how we identify species in nature using the taxonomists' tool, diagnosis.

If we examine how useful fixed characters are in delineating species, we find that the paradox is an artifact of the requirement not only that individuals belong to no more than one species but that this placement be immutable through time. The heart of Woodger's paradox is that species are biologically distinct in some unspecified sense. But if this unspecified sense is simply one of recognition, the paradox goes away, because a species determination for a given individual is temporally dependent: the inclusion of an individual in a particular species occurs as a result of genetic fixation events that make one entity reproductively distinct from others. And classification of individual organisms into entity A or B most certainly can change with time, as character fixation events might come and go as a result of re-established genetic contact. Woodger's vision of speciation is of a temporal boundary drawn between two successive generations, and this is where the problem occurs. Character fixation is an implication of the phylogenetic criterion that "speciation" is an instantaneous event, corresponding to character fixation and coinciding with the death of the last individual with character states that conflict with diagnosis. This particular view of life evolving is inconsistent with Woodger's premise that speciation occurs between two successive generations. Woodger's suggestion, that speciation is more or less instantaneous, is an apparent outcome of the branching genealogy that Darwin used as his only figure in the *Origin*. Which brings us back to branching diagrams and tracing the history of humans.

MAKE THREE WISHES

Earlier, in our discussion of the "line of death" at which species form, we stated pretty strongly that trees are not a valid way to think about relationships among individuals, like you and us, who all belong to the same species. Yet we have all seen trees drawn to identify the patterns of human movement across our planet. In addition, we all have heard of DNA-based "Adams" and "Eves" hypothesized on the basis of trees. The reason that trees and human ancestry have gone hand in hand is the result of a strange happenstance. If we were given three wishes to order up a tool for tracing the history of our species, we would almost certainly first wish for something that inherently carried the history of our species. One tool that scientists have settled on is DNA, which, because of the way it is inherited, does indeed encode our history. But, as we pointed out earlier, the history of genes can become jumbled because of recombination and because one gene might have a completely different history from another. We might thus use our second wish for a simpler tool than the nuclear DNA that carries the main "blueprint" for constructing our bodies. And if this tool simply followed female history, we might also wish for a third tool that followed male history. Luckily, such tools exist, and they exist because of the peculiar nature of our biology.

For over three decades now, the mitochondrial genome has been used for tracing female ancestry. In most animals, the mitochondrial genome is a small circular piece of DNA containing sixteen thousand G, A, T, and C bases. It resides not in the cell's nucleus but coiled up in the mitochondria, the so-called powerhouses of the cell, and it incorporates thirteen protein-coding genes and several genes that make RNA. During fertilization the mitochondria from the male sperm (sperm are mostly just nuclei but need mitochondria to produce the energy that drives their tails for movement) don't usually make it into the egg. As a result, male mitochondria are essentially a genetic dead-end. Female mitochondria, on the other hand, are passed on to all offspring regardless of sex. Males need and possess mitochondria, but they do not pass on the circular genome residing in this organelle. Effectively, then, mitochondrial DNA (mtDNA) maps female lineages very precisely. And it isn't just this mode of inheritance that makes mitochondrial DNA useful for following females, for since mitochondrial genomes from two different individuals rarely recombine with each other, their pattern of inheritance is clonal. This latter aspect of mtDNA is as important as maternal transmission in making this molecule the powerhouse tool for studying female movement around the world that it has become.

But let's have equal time here, and for males, an equally well-suited tool has been used. This time it is the Y chromosome, the hallmark of the male sex. The transmission of the Y chromosome is strictly from father to son. Females who can reproduce do not have a Y chromosome and instead usually have two X chromosomes. X chromosomes are loaded with genes, over two thousand of them. In contrast, there are probably fewer than fifty genes on the human Y chromosome. Most males have one X chromosome (they get this from their mothers) and a single Y chromosome (which they get from their fathers). Gentlemen, because of this, not only are we mitochondrial dead-ends, but we are also X-challenged, as any female old enough to interact with us will attest. But again, our single Y isn't the whole story as to why this chromosome is such a great tool for following males. Because the Y chromosome doesn't have a chromosome to pair with (it only weakly recognizes the X chromosome), it effectively does not recombine. So it, too, evolves in a clonal fashion, much like the mtDNA genome. One couldn't have thought of better tools for tracing human ancestry than the mitochondrial DNA and the Y chromosome, and both are widely used nowadays as the major tracers of human ancestry.

Currently, it is a piece of cake to sequence the entire mitochondrial genome (sixteen thousand bases) of an individual human. It is also really easy to get sequence for one-fifth to one-half of the genes on the Y chromosome. Large amounts of data can thus be collected for these two markers, and these allow us to follow human migration patterns.

The way this works is to obtain the sequences for a large number of individuals from diverse geographic origins, and then to construct their mtDNA genealogy based on sequence similarities and differences. The resulting trees will show how closely related females from different geographic regions are. The trees then are interpreted as "roadmaps" showing the spread of females across the planet. In a similar manner, the Y chromosome data are collected and used to construct a Y-DNA genealogy. This can be interpreted in the same way as the mtDNA information, but it reveals how males spread across the planet. Again, we face problems when we attempt to use these trees and their interpretation to say things about an individual's ancestry; and there is no doubt that these approaches are best used bearing in mind that we are looking at the evolution of mtDNA and Y chromosomes rather than of individuals. Since genealogy has often been used to replace "race" in modern biology, we will return later to the problem of interpreting trees and how this relates to race. The substitution of genealogy for race actually requires very close scrutiny.

HUMANS AS ANOTHER ORGANISM TO SYSTEMATIZE?

All of the techniques and approaches we have discussed in this chapter point us toward a simple conclusion. Using the most widely accepted definition of species, and using the most precise criteria for species delimitation, we are left with the conclusion that there is but one species of human being currently in existence on this planet. The taxonomy of our species, *Homo sapiens,* is clear. All modern DNA-based tools, such as mtDNA and Y chromosomal DNA, point to this conclusion, and in addition, if we tried to DNA barcode our species, we could easily be barcoded together as distinct from other species on this planet, most notably from our close relatives the bonobos and chimpanzees. However, because of current broad and rampant interbreeding between populations of diverse geographical origin, and the fact that reintegration of once-isolated human populations has been occurring now for several centuries and even millennia, the attempt to delimit smaller units within our species is flawed. There are two reasons for this. The first is empirical, and it rests on the fact that no aggregate of humans based on geography or outward appearance would give up any diagnostics at all. The second is philosophical and lies at the heart of taxonomy. In order for taxonomy to work, it needs to be hypothesis-based. This requires that hypotheses be erected about species status that can then be tested using empirical data. There are *no* valid hypotheses of species existence that we can think of for testing within our species. In other words, there are no smaller aggregates that invite testing as species within our broader human population because of the rampant interbreeding we see on our planet today. Reintegration is smoothly underway, and if things keep going the way they are now, it will be beautifully irreversible. As remarkable as it might sound to humans with a keen ear for nuance, all humans together are the smallest diagnosable aggregate of individuals that can be conjured up to be tested for species existence.

We have spent a great deal of time on what species are and how they are recognized taxonomically. This is because we need this basic background to discuss the weaker taxonomic levels of "subspecies" and "race." While there are strong objective and operational criteria for the recognition of species, the criteria for giving an entity or population the subspecies epithet are much weaker and more subjective. Since subspecies and races can in many ways be equated, it follows that the epithet "race" is an equally subjective and weak taxonomic delimiter—if not more so. Also, if we say that the epithet "race" denotes a level that is yet less organized than subspecies, we are faced with an even bigger subjectivity problem in using race as a term.

What, then, are subspecies and races in a biological context? We propose that both of these well-worn taxonomic levels are, simply put, hypotheses of species existence that have not been tested. When a biologist suggests that he can recognize a population (or aggregate) as a subspecies, he is making a claim that the population might be distinct from other closely related populations in the same species. He is, in other words, hypothesizing that what he has designated a subspecies might be a species. Some biologists might argue that subspecies designations are useful, because they recognize something interesting or special about an aggregate of individuals within a species. We agree completely but point out that, whatever their inherent interest, what makes subspecies special to the systematist is merely that they might be species. There is, to us, no other systematic use for the epithet "subspecies" than in denoting a hypothesis.

Using such a hypothesis is rather like flagging the population for further study. What the subspecies epithet does is to say, "Hey, wait a minute, don't forget about me in your next grant proposal." Likewise, while "race" has been used in a rather weaker sense to denote moderately differentiated populations as somehow "special," it is also no more than a hypothesis of species existence. And as we have already suggested, there are no robust hypotheses of aggregation of humans, other than that we are all one single aggregate. In essence, we have indeed tested hypotheses of subspecies existence in humans and of the existence of race in humans, and these tests have been falsified, over and over again. As a result, erecting any hypothesis of race or of subspecies within our species seems to us to be misguided. It is time to stop hypothesizing about the existence of more than a single species or aggregate of humans on the planet today and to move on to other, more important, aspects of our biology and systematics. Yet many biologists still can't resist using the term "race," thus implicitly projecting the idea that it is a biologically valid notion. We will return later to the artificiality of the problems raised by this reluctance to move on.

CHAPTER 3

HUMAN EVOLUTION AND DISPERSAL

THE ROOTS OF the human family go back some seven million years, or maybe a bit more. The order Primates to which we belong is first documented around the time the dinosaurs became extinct, some sixty-five million years ago, though early representatives might already have been around for some time. The first known mammal relative lived back in the Carboniferous, well over three hundred million years ago. In contrast, we human beings are mere parvenus. The fossil record and molecular comparisons agree that the human species is of remarkably recent origin and that it dispersed worldwide even more recently. Viewed in evolutionary perspective, those features that we recognize as "racial" were acquired only yesterday, mere seconds before midnight on the evolutionary clock, as it were; and what's more, unlike furry skin, grasping hands, or upright posture, as intraspecific variations they are entirely epiphenomenal. In this chapter we will attempt to put them in historical perspective by briefly looking at the evolutionary history of the human family and at that fleeting moment in time that witnessed the origin and spread of *Homo sapiens*.

As documented by fossils (principally the petrified bones and teeth of ancient animals preserved in sedimentary rocks), the story of our zoological family Hominidae (which you will often see alternatively classified as the subfamily Homininae; it's of no consequence here) goes back to almost seven million years ago. This was the time around which the first creatures emerged that were more closely related to *Homo sapiens* than to any of the apes. The earliest known hominids, all of which date to the period between about seven and four million years ago, come from sites in Africa—and they make up a pretty motley bunch. What all mainly have in common is the claim, made by their describers for a variety of different reasons, that each was an upright biped, walking on two legs when on the ground—where it seems they actually didn't spend all, or even most, of the time. Descended from arboreal ancestors, the early hominids appear to have lived at a moment when drying and increased seasonality of the African climate was beginning to shrink the continent's formerly monolithic forest cover, but their remains were all found alongside the remains of animals typical of fairly forested conditions, or at least of a mosaic of habitats ranging from forest to grassy woodlands. In other words, our early precursors of this period may have come to the ground, but they hadn't yet abandoned the trees as their successors would ultimately do.

IN THE BEGINNING

The earliest of the putative early hominids is a form from Chad imposingly named *Sahelanthropus tchadensis*. Known principally from a skull and some other bits thought to be close to seven million years old, *Sahelanthropus* is a primate with a small (ape-sized) braincase but with a flattish face housing teeth that are not at all ape-like. In particular, the canines seem relatively reduced so that they do not project much beyond the level of the other teeth, and the upper canine does not sharpen itself on the lower premolar behind. What's more, the molar teeth are squarish, and, in contrast to those of apes, they are coated with quite thick enamel, usually taken as a sign that relatively tough foods (maybe the tubers of plants that typically grew in more open areas) were being chewed, in contrast to the soft forest fruits favored by ancestral apes. Over the years much has been made of the significance of canine reduction—it has been putatively associated, for example, with changes in social organization, possibly in the direction of pair-bonding between males and females, making the reproductive unit a single adult couple. And once you have made that association, it doesn't take much to imagine bipedal males foraging far and wide and filling their arms with food that was transported back to females and infants, reinforcing the pair bond and allowing females to reproduce more frequently, with the rest being history. Well, maybe. But it's a bit of a stretch, and it's at least as plausible to conclude that canine reduction "unlocked" the lower from the upper jaw, allowing side-to-side grinding of those tougher foodstuffs. As for bipedality, the single known skull of *Sahelanthropus* is crushed but sports a foramen magnum (the large hole through which the spinal cord exits the brain) that lies somewhat under the braincase, suggesting that the skull was balanced atop a vertical spine. That the foramen magnum was indeed tucked under the skull seems to have been confirmed by a recent "virtual" reconstruction of the distorted original (accomplished entirely inside a computer).

Another pretender to early hominid status is *Orrorin tugenensis*, a six-million-year-old form from northern Kenya. The handful of partial leg bones from which this primate is mainly known are not entirely diagnostic of bipedality but are consistent with it, and the molars are comfortably broad and thick-enameled, as one would expect to find in a precursor to later members of the family. The final entrant in the "very early hominid" category is *Ardipithecus*, a genus reported from Ethiopia beginning around 5.8 million years ago. Its best-known representative is the 4.4-million-year-old *Ar. ramidus*, which is a very odd bird. A badly crushed but remarkably complete skeleton of this form has a reduced canine complex and, as reconstructed, a fairly forwardly placed foramen magnum, but its body skeleton shows few features that might be expected of an upright walker on the ground. For a start, its feet have a strongly divergent great toe, in contrast to all later hominids with preserved feet, in which this toe is in line with the others. The foot of "Ardi" is a grasping foot, not a walking one. Yet its forelimbs are said not to possess any features relating to suspension of its body weight in the trees—something that would have been virtually obligatory for an arboreal form that weighed well over a hundred pounds. One feature of the badly crushed pelvis is said to support uprightness, but it is not a feature that is unique to hominids. All in all, Ardi is a very peculiar primate indeed. And it's certainly not a direct human precursor, for, as we will shortly see, it is very little older than a much better claimant for that title.

The fact that these putative early hominids are so oddly assorted can be interpreted in a variety of ways, but at the very least this diversity makes a powerful general point about our family: from the very beginning the story of our precursors was one of energetic exploration of the possibilities inherent in being hominid. We are, in other words, very much misled if we regard the status of *Homo sapiens* as the only hominid in the world today as anything like "normal." It may seem natural to take what we are familiar with as typical for human evolutionary history; and if things were really that way, we would certainly want to reconstruct our past by projecting our lonely species way back in time in a single unbroken lineage. Indeed, as we saw in looking at the Synthesis, it is not so long since this was more or less standard operating procedure in paleoanthropology. But by now it's clear that this linear scheme furnishes us with a highly inaccurate image of our biological past. Instead, as the varied early hominids

show, the picture was from the very start one of vigorous evolutionary experimenta-
tion, as one species of early hominid after another was thrown up by the evolutionary
process and pitchforked into the ecological arena to become extinct or give rise to
descendant species. Nothing linear here, as a glance at figure 3 will reveal.

THE "AUSTRALOPITHS"

At 4.2 million years ago, in northern Kenya, we find the first evidence of a hominid
species called *Australopithecus anamensis*. This is the first member of our family whose
fossil leg and foot bones speak directly of upright bipedality. Its jaws and teeth were also
comfortingly similar to the next-in-time *Australopithecus afarensis*, a hominid whose
fossils are widely known in eastern Africa between about 3.6 and 3.0 million years ago.
Most famously represented by the 3.2-million-year-old partial skeleton "Lucy," from
Hadar in Ethiopia, A. *afarensis* is the best known of the early hominid species. Abundant fossils demonstrate that members of this diminutive species (females stood three
and a half feet tall; males were up to a foot taller) were indeed upright walkers on the
ground. But with their rather short legs and flaring hips it's clear that they didn't walk
around exactly as we do, and they also retained a lot of features that would have helped
them climb quite adroitly in the trees. Indeed, they probably still largely depended on
trees for shelter and sustenance.

    Above the neck, these "australopiths" remained substantially ape-like, with large
projecting faces and small cranial vaults that contained brains no larger than you'd
expect to find in an ape of the same body size. For this reason, many paleoanthropologists have come to regard them essentially as "bipedal apes," primates that came
intermittently to the ground less because the ancestral forest habitat was disappearing
entirely (the ancestral great apes stayed in its remnants, after all), than to exploit the
new resources offered by grassy woodland environments.

    These environments would have presented an unaccustomed new suite of dan-
gers, among which would have been the terrestrial predators that then existed in Africa
in even more fearsome array than they do now. Beginning in the late 1940s Raymond
Dart, the describer of the first-discovered "australopith," popularized the notion that
the early bipeds were vicious "killer apes" and "murderers and flesh-hunters," setting
the scene early on for the mighty hunters and serial killers of today. Were this view correct, the social groups of the bipedal apes would almost certainly have been smallish,

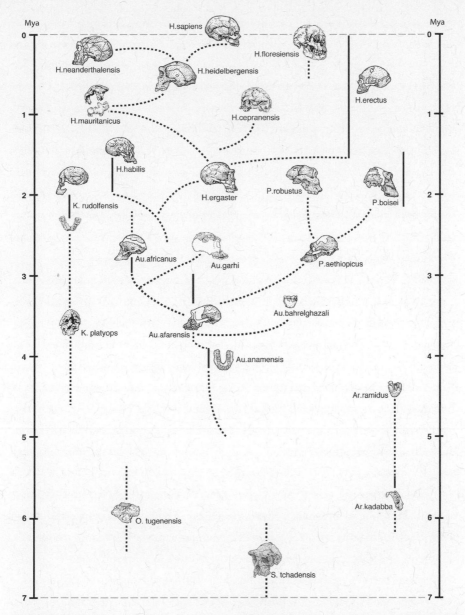

FIGURE 3. *This diagram shows the distributions in time of the various generally recognized species of the family Hominidae over the past seven million years or so and some potential links between them. The primary message here is the strong signal of species diversity at virtually all points on this time chart— except today.* ©*Ian Tattersall.*

as is typical for predators, but the reality was probably very different. The slow-moving and small-bodied bipeds were highly vulnerable to predation away from the trees, and it is quite likely that, like other prey species, they sought safety in numbers, much as our baboon and macaque relatives do in relatively open environments today. Chemical signals in their bones and teeth suggest that they may themselves have been eating small browsing prey, such as hyraxes, but it is likely that most or at least much of their sustenance came from plants. With their small canine teeth and diminutive bodies they were almost certainly not great aggressors, and individually they would have been more or less defenseless against predators. Within the large groups that such realities would have made almost inevitable, social relationships were undoubtedly intricate and subtle, but there's no reason to believe that the early hominids were cognitively any more sophisticated than the great apes are now.

This is not to demean the australopiths in any way; apes are cognitively very complex beings, and hardly a month passes in which primatologists do not discover that they perform yet another behavior that we had thought was unique to us. Yet apes clearly do not process information about the world around them in quite the way we do, and based on his keen observations of ape behavior the cognitive scientist Daniel Povinelli hazarded not long ago that if we were to board a time machine in search of early hominids we would debark to find "intelligent, thinking creatures who deftly attend to and learn about the regularities that unfold in the world around them. But . . . do not reason about unobservable things: they have no ideas about the 'mind,' no notion of 'causation.'"

Hominids of the general australopith kind flourished widely in Africa between about 4 and 1.4 million years ago, branching out into a wide variety of species, some of which had quite distinctive morphologies. Again, as with the very earliest hominids, these creatures were displaying a tendency to diversify in different places, a proclivity that has been the hallmark of our family ever since. There is nothing surprising about this; indeed, it is typically what happens in evolution when a significant structural innovation opens up new ecological possibilities to members of an established group. The process even has a name: "adaptive radiation." So it was with the first bipeds, but although the emergence of the australopiths clearly represented a vital step in our direction, it is clearly misleading to view them as somehow "transitional" between their arboreal ancestors and us. For theirs was a stable and successful body form (and presumably general life way) that continued to serve them well for millions of years, even as a succession of australopith species stepped on to, and exited, the evolutionary stage.

THE EARLIEST STONE TOOLMAKERS

Human beings are unique in having an archaeological record—a material documentation of at least certain aspects of past cultural behavior. Well, almost unique: primatologists have recently found some stone "anvils" on which West African chimpanzees cracked nuts several thousand years ago. The human archaeological record begins with the first recognizable stone tools found at East African sites up to about 2.5 million years old, mostly places on the landscape where animals—probably mainly scavenged—were butchered, using sharp stone flakes knocked off one piece of rock using another. Intriguingly, evidence has recently been reported of animal bones from the site of Dikika, in Ethiopia, some 3.4 million years old, that bear cut-marks of the kind that stone tools make when used for butchery. No stone tools are known from the site, but remains of *Australopithecus afarensis* were found nearby. However far back in time the practice of dismembering animal carcasses with stone tools goes, one thing is certain: it provides, at last, clear evidence for a cognitive leap beyond the ape condition. For, even with intensive training, no modern ape has been able to fully grasp the idea of hitting one rock with another at precisely the force and angle necessary to detach a usable flake. What's more, the early tool makers showed considerable powers of anticipation, carrying suitable lumps of stone around with them for many kilometers before making them into tools as needed. Yet it seems pretty clear now that the early stone tool makers were physically archaic: not significantly different anatomically from the creatures that had preceded them for many millions of years. Right from the very beginning, then, we see evidence for what was to become a durable pattern in the hominid record: the decoupling of biological from cultural innovations.

The invention of the cutting flake was probably the most fateful and important human cultural innovation ever, and it was followed by a long period of monotony in stone tool production. For over a million years there was no significant technological advance, and these basic tools continued to be made even after a radically new kind of hominid came on the scene. Best known from the 1.6-million-year-old "Turkana Boy" skeleton from northern Kenya, this new kind of hominid is often known as *Homo ergaster* (though sometimes as *Homo erectus*), and it is the first member of our lineage to boast more or less modern body size and proportions. The oldest fossils usually attributed to *Homo ergaster* are a little under two million years old, and the youngest are about 1.5 million. The various skulls often assigned to this species make a pretty motley assortment, and if several taxa are not in fact involved here, *Homo ergaster* was a very highly variable species indeed.

Whatever the taxonomic case, *Homo ergaster* was very different from its predecessors: a tall, slender, long-legged strider, well adapted to a life on the open, shadeless savanna beyond the forest fringes to which its predecessors had been confined. This was a radically new kind of hominid, with enormous and intimidating new possibilities open to it. Many believe that the new body form not only made possible but dictated a new lifestyle, one that involved obligate hunting—because not that many other resources besides plant tubers and the occasional lizard would have been open to hominids out there on the savanna. For all of our sedentary ways today, members of the genus *Homo* are incomparable long-distance walkers and runners. They may not be fast, but provided access to water they can keep going almost indefinitely; and while they cannot outrun their prey, they can exhaust it. Such sophisticated practices as wounding a prey animal with a poisoned throwing spear, and then tracking it through indirect signs, were still far in the future 1.6 million years ago; but a doggedly determined hominid, capable of sweating through its naked skin to keep cool, could still have chased an antelope out in the noonday sun—and have continued chasing it, to the point of the animal's exhaustion. *Homo ergaster* may have been doing it the hard way, but here was an unprecedented *kind* of hunter.

Yet for all of its new features, both behavioral and physical, *Homo ergaster* possessed a brain only modestly larger than that of the bipedal apes—and still only about half the size of our own today. And despite the fact that these creatures must have experienced a lifestyle very different from that of the woodland-bound australopiths, we see no significant change in the stone tool record until well after they had appeared on the scene. Within the million-year period of technological stasis, there is evidence for considerable local variation in the kind of environmental challenge faced by the tool makers, whoever they were, and evidently they responded to these challenges by using the old kinds of tool in new ways. But at about 1.5 million years ago, when *Homo ergaster* had already been on the scene for a longish time, a new kind of stone tool finally showed up. This was the handaxe, a largish implement, usually about eight inches long, that was elaborately fashioned to a predetermined symmetrical teardrop shape, using multiple blows to both sides. Here we see a radical departure in how stone tools were perceived by their makers. Earlier toolmakers had simply smashed off a flake, whose exact appearance mattered little as long as it had a cutting edge. In contrast, the handaxe provides us with excellent evidence of a new and more complex way of looking at and thinking about the world and the materials it offered. The making of enormous quantities of handaxes to a set "mental template" that must have existed in the toolmaker's mind before knapping started, surely implies a major advance in cognitive complexity. Sadly, though, we don't know what the wider implications of this advance were; there is no evidence to tell us what this new way of envisioning possibilities meant for the handaxe makers' wider perception of the world or, indeed, of themselves.

EXODUS

Shortly after the appearance of the new hominid body form in eastern Africa, and well before the invention of the handaxe, early *Homo* spread out of Africa, in a process of demographic expansion that was certainly facilitated by the new striding body form and probably also by environmental changes that made the Eurasian continent easier for hominids to penetrate. By 1.8 million years ago hominids were in the Caucasus, between the Black and Caspian Seas, and probably not long thereafter they had already reached eastern Asia, where the endemic form *Homo erectus* rapidly evolved. By 1.4 million years or so, hominids had even penetrated the climatically more difficult European peninsula. Once more, the evidence strongly suggests local divergence and differentiation in each of the regions entered, just as one might expect for any successful and widespread mammal group. Sporadic fossil finds in Africa suggest that differentiation was also continuing in the parent continent.

The first cosmopolitan species in our genus is *Homo heidelbergensis,* so named because the first example was discovered in a gravel pit not far from Heidelberg in Germany. This specimen has recently been dated to slightly over six hundred thousand years ago, but the species itself seems to have originated in Africa, where an Ethiopian cranium is known of similar age. Later examples come from sites in Europe and China as well as from Africa, the youngest of them perhaps only a couple of hundred thousand years old. Here was a hominid that was very much like us, albeit very robustly built and with a brain only just within the lower end of our size range.

Despite the rather small sample of reasonably complete crania, *Homo heidelbergensis* clearly shows distinctive variants, though whether these variants are organized regionally is a bit less clear. With a heavy face that was still hafted in front of the brain case, rather than tucked beneath it as ours is, *Homo heidelbergensis* initially at least made very crude stone tools. Still, it was in the tenure of this species that fire began to be regularly domesticated (the earliest definite hearths known go back to about eight hundred thousand years, but it was only much more recently that fire use seems to have become common). This is also the time from which the first shelters are known, and from which the earliest wooden throwing spears survive. Clearly, hominids at around four hundred thousand years ago were using the resources around them in newly complex ways; but alas, once again we know nothing about the broader cognitive correlates of such developments.

It was also within the tenure of *Homo heidelbergensis* that the next major innovation in stone toolmaking was introduced. This was the "prepared-core tool," made by flaking a stone "core" on all sides until a single blow would detach a more or less finished tool. Again, we have here prima facie evidence of increasingly complex cognition, but we arrive at this frustratingly general conclusion in the absence of any clear idea what exactly "more complex cognition" actually meant for the lives of the hominids concerned and for their subjective experience of them. What is pretty clear, though, is that even late in the *Homo heidelbergensis* record there is no unequivocal evidence for the production of symbolic objects. This is important, because it seems to be the facility to create and manipulate mental symbols that sets modern *Homo sapiens* off from every other kind of hominid that has ever existed. Our capacity to dissect the world around us into a vocabulary of discrete symbols, and then to recombine them to create new possibilities, seems to be the key to our ability to literally remake the world in our heads. All other species appear to react, with greater or lesser sophistication, more or less directly to the stimuli that nature presents to them. In contrast, and uniquely, we modern *Homo sapiens* live largely in worlds of our own making. And, to judge from the material record it left behind, *Homo heidelbergensis* did not.

This also seems to have been true for *Homo neanderthalensis*, the best-known of the extinct hominids and quite possibly the most skilled practitioner ever of the prepared-core technique. This species shared with *Homo sapiens* an ancestor that must have lived in Africa at some time well over half a million years ago, and it was the issue of a long separate evolutionary trajectory in the remote fastnesses of Europe. Sadly, while we have good fossil evidence for Neanderthal antecedents in Europe at about the half-million-year mark, we have as yet nothing equivalent for *Homo sapiens* in Africa. Most strikingly, *Homo neanderthalensis* shared with *Homo sapiens* the large brain that distinguishes us as a species today. The increase in brain size apparently occurred independently in the Neanderthal and *Homo sapiens* lineages, albeit from a fairly advanced starting point.

First recognizable at about two hundred thousand years ago, and like us remaining a distinctive entity even while it apparently differentiated in different parts of its range, *Homo neanderthalensis* seems to have dominated the European and western Asian landscape until about forty thousand years ago, when the first *Homo sapiens* émigrés from Africa began trickling into its heartland. Within ten millennia of this event the Neanderthals were entirely gone, having been displaced as a result of direct conflict, indirect economic competition, or more likely a combination of the two.

There is good molecular evidence to support the conclusions both that, while the Neanderthal populations, like ours, differed in minor features from time to time and place to place, the Neanderthals were a discrete genetic entity and that they made no significant genetic contribution to today's European population. There may have been some physical interchange between the two species, but neither left its physical stamp upon the other, and there was evidently minimal genetic intermingling of the populations, as the recently announced draft Neanderthal genome appears to show.

The resulting individuality of the Neanderthals makes them an instructive comparison to the *Homo sapiens* who invaded their territory some forty thousand years ago and shared it with them for several millennia. In contrast to the essentially symbol-free lives of the Neanderthals, those of the intruders were drenched in symbolism. The incoming modern people were the "Cro-Magnons," who left us the astonishing legacy of Franco-Cantabrian Ice Age cave art at such legendary sites as Lascaux, Chauvet, and Altamira. The archaeological record they left behind also contains evidence of their symbolic mental processes: musical instruments, decorated artifacts, sculpture, systems of notation, and even ceramics.

One must admire the achievements of the Neanderthals, who survived a very long time on some very difficult landscapes. And indeed, perhaps the Neanderthals exhibited the ultimate in what a purely intuitive, nonsymbolic intelligence can achieve. But when you contrast the material evidence they left behind with that of the Cro-Magnons, it is hard not to conclude that we are looking at the echoes of entirely different beings. The Cro-Magnons were *us*; the Neanderthals were not. They were among the last of the nonsymbolic old breed.

THE ORIGIN OF MODERN *HOMO SAPIENS*

The Cro-Magnons brought a fully formed modern sensibility into Europe. This unprecedented cognitive quality had originated elsewhere, and that place was almost certainly somewhere in Africa, the continent in which our very peculiar modern anatomy also evolved. It is worth noting that our anatomical modifications reverberate comprehensively from head to foot, as is evident from figure 4 comparing a modern human skeleton with that of a reconstructed Neanderthal. Particularly in the form of the body skeleton, it is the latter that better represents the primitive form for the genus *Homo*.

The earliest suggestion we have of beings on the planet who looked pretty much like us in their bony structure comes from Kibish, in southern Ethiopia, in rocks now dated to about 195 thousand years ago. Two hominid crania have been found in the Kibish deposits. One of them looks rather archaic, but the other one, pieced together from fragments, looks remarkably like us, with a voluminous cranial vault and a small, retracted face. Oddly, it does not have two of the most striking of our cranial features: a diagonal bony groove above each eye that produces a unique configuration of the eyebrow area, and a chin with the specific form that ours has. Nonetheless, the strong similarities between the Kibish cranium and ours compel its attribution to *Homo sapiens,* and the features it has in common with us are all the more striking because they are pretty much unanticipated in the known fossil record. In contrast to the Neanderthals, who have recognizable fossil precursors dating back to half a million years ago or more, *Homo sapiens* shows up pretty much unannounced. Doubtless this "sudden" appearance of substantially modern anatomy is at least in part an artifact of a relatively poor African fossil record for the period in question, but it nonetheless underlines the remarkable distinctiveness of *Homo sapiens,* whose origins plausibly lie in a single innovation in gene regulation or expression that had ramifying effects throughout the body.

Let's digress briefly and return to Allan Wilson's body of work to understand the role of simple regulatory changes resulting in the vast anatomical and behavioral differences we see between humans and our closest living relatives the chimpanzees. In the 1970s Wilson and Mary Claire King, a University of Washington biologist, noticed

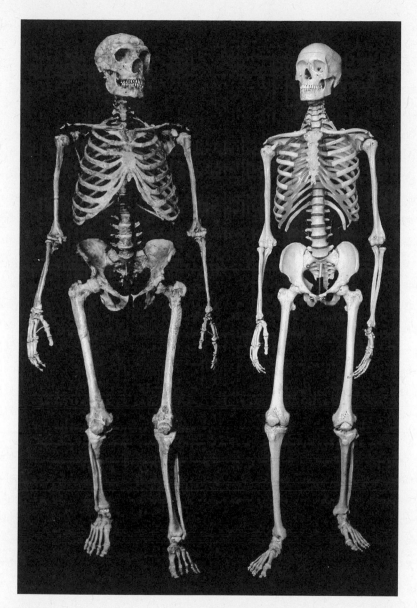

FIGURE 4. *Contrast between a reconstructed composite Neanderthal skeleton (on the left) and that of a modern human of similar stature. Note particularly the dramatic differences in the proportions of the thorax and the pelvic girdle (with the Neanderthal probably displaying more or less the primitive condition for the genus* Homo: Homo sapiens *emerges from this comparison as very highly derived). Photo by Ken Mowbray, ©Ian Tattersall.*

that the anatomical differences between organisms were much greater than the differences seen between their protein-coding gene sequences. In fact, Wilson and King showed that the difference in the average protein-coding gene sequences of chimps and modern humans was about 1 percent. In other words, the proteins that we use in our day-to-day biology are nearly identical to those that chimpanzees and bonobos use. Wilson and King then reasoned that this small amount of change in protein sequences could not by itself account for the huge differences seen in the anatomy of these organisms. Because genes also have regions associated with them that act as regulators of the activity and amount of the proteins they code for, they turned their attention to these regulatory regions as the arbiters of large anatomical and behavioral changes. They proposed that slight changes in regulatory genes would have a much greater effect on anatomy and behavior than would the very slight changes they observed in the structure of proteins. Now, three decades later, the genomic revolution has bolstered their initial hypothesis, as the sequence difference in coding regions of genes between chimps and humans is reported as between 1.5 percent and 1 percent.

Genomics has since allowed scientists to take the idea a step further. By looking at the rate of gene expression of many genes at once, in many different tissues such as different regions of the brain, scientists have been able to show clearly that gene products in chimpanzees and humans are regulated quite differently. For instance, Wolfgang Enard, Monica Udin, Mario Cáceres, and their colleagues have shown that since their common ancestor some 80 to 90 percent of a specific set of genes have achieved increased expression levels in the cerebral cortex of humans, compared to an increase of only about 15 to 20 percent for chimpanzees. In another study, Svante Pääbo and his colleagues at the Max Planck Institute in Germany examined more localized regions of the brain such as the cerebral cortex, the caudate nucleus, and the cerebellum. They showed that these three regions differ in gene expression patterns from each other, within species, and between humans and chimpanzees, and found that "10% of genes differ in their expression in at least one region of the brain." Finally, Kevin White and colleagues have determined that some of the expression differences are due to a specific kind of regulatory genes called transcription factors. These proteins regulate the expression of other genes in the genome, functioning rather like switches. Often, a slight change in one transcription factor will alter the expression of many genes at once. These early studies of gene expression in primate brains indicate that substantial changes in gene expression have intervened between us and one of our closest living relatives, the chimpanzee, and they point to changes in gene regulation as the potential source for the vast differences in anatomy we see between us and our closest living relatives.

Molecular geneticists have thus furnished us with a potential mechanism for the apparently rather sudden appearance on Earth of hominids who looked pretty much the way we do. The Kibish hominid has most of the necessary hallmarks for membership in our species, at almost two hundred thousand years ago; and confirmation of the arrival of *Homo sapiens* is soon forthcoming in the form of a similarly voluminous 160,000-year-old cranium recently discovered at a place called Herto, also in Ethiopia. In its anatomical features this specimen is said to fall marginally outside the range of variation seen in *Homo sapiens* today, and it has consequently been placed in its own subspecies *Homo sapiens idaltu*. Still, it is worth recalling that subspecies are at best an iffy category when you are dealing with fossils (or with living humans, for that matter) and that the principal importance of this fossil is in demonstrating conclusively that the basic modern human anatomy had become established in the period not long after two hundred thousand years ago. Interestingly, stone tools found in the same deposits as both Ethiopian specimens are pretty unimpressive; the few examples reported along with the Omo Kibish 1 cranium have been described as "unremarkable," while those from the same deposits as the Herto cranium are notably archaic and include some of the latest recorded African handaxes. They also embrace some Middle Stone Age elements, prepared-core tools roughly equivalent to the productions of Neanderthals. So we find an old pattern repeated again: here is evidence of a physically new kind of hominid (indeed, radically so), in the absence of any suggestion whatever of any technological advance. This might on the face of it seem a bit counterintuitive, but in fact it's not unexpected: any technological innovation has to start with an individual, who is necessarily a member of an established species, however much Woodger might have agonized about it.

Various other finds in Africa from the period between two hundred and one hundred thousand years ago also suggest that *Homo sapiens* had become established in this continent well within this time. But the first hint of humans of modern bony anatomy outside Africa do not come until about ninety-three thousand years ago, the age of archaeological levels at the Israeli site of Jebel Qafzeh that have produced a couple of unquestionably modern skeletons. Interestingly, though, these remains are associated with a stone tool kit that was functionally identical to that made by the

Neanderthals who occupied the same area both before and after that time. Prior to the eventual disappearance of the Neanderthals at about forty thousand years ago, there was evidently a long period during which neither species managed to dislodge the other definitively from this region. Significantly, during this long period of coexistence (or possibly of time-sharing; there are no sites at which the species co-occur) there is little to suggest any cognitive difference between Neanderthals and anatomically modern humans. The early Levantine *Homo sapiens* were doing business in the traditional way, and there are no substantial reasons to believe that they were symbolically processing information in their heads.

The first stirrings of symbolic activity in the archaeological record come, like the first intimations of anatomical modernity, from Africa. The very earliest suggestion of a modern behavioral style comes from the sites of Klasies River Mouth on Africa's southern coast, where occupation levels dated to over one hundred thousand years ago are said to show a symbolic organization of the living space. A little farther west along the same coast are the sites of Pinnacle Point and Blombos Cave. The main archaeological levels at both of these sites are attributed to the Middle Stone Age, recognized by a set of stoneworking industries that are roughly equivalent to the Middle Paleolithic of the Neanderthals in Europe and the Levant. However, it is becoming ever clearer that something more complex than we see in the Middle Paleolithic was astir at some Middle Stone Age sites in Africa. Bone harpoon points of a complexity not seen in Europe until long after the Cro-Magnons had arrived are known from a site in the eastern Congo dated to between sixty and ninety thousand years ago, and one especially suggestive technological development has been recently reported from levels at Pinnacle Point that are over seventy thousand years old. There, early humans began heat-treating silcrete, a form of cemented silica that is normally a rather poor material for making tools. However, when carefully heated and cooled in a very long and complex process, it changes color and transforms into a substance from which excellent tools can be made. There are not many aspects of Old Stone Age technology that can be taken as presumptive evidence for symbolic mental processes on the part of the users—Neanderthals made beautiful stone tools without apparently being symbolic—but here we pretty plainly have just such evidence. The multistage heat-treatment process is so complex, and so dependent on careful planning, that few would dispute that it had to be the product of symbolic minds. Sadly, there are no fossils from Pinnacle Point that would tell us just who those symbolic hominids were, but the betting has to be that they were anatomically modern *Homo sapiens*.

For overtly symbolic objects, comparable to some of what the extraordinary Cro-Magnons produced, we have to move yet farther westward, to Blombos Cave, close to Africa's southern tip. There, in deposits about seventy-five thousand years old, two smoothed ochre plaques bear distinctive geometric engravings, the earliest direct evidence acceptable to most observers of a symbolic mind at work. Less direct evidence comes from adjacent deposits at Blombos in the form of small shells apparently pierced for stringing, hence body decoration. In all documented human societies, body ornamentation is heavily loaded with connotations of status and identity and is widely accepted as prima facie evidence of modern human cognition. Since the Blombos find was announced in 2004, reports have come in from other areas of Africa, and even from the Levant, of shell "beads" of similar or even greater antiquity. Even earlier in time, at both Blombos and Pinnacle Point, evidence of the demanding pursuit of shellfishing that may date back to as much as 140 thousand years ago has been said to indicate complex mental processes very far back in time; but since this activity was also indulged in by nonsymbolic Neanderthals in Gibraltar, it is probably not a reliable proxy for symbolic cognition.

SO WHAT HAPPENED?

After around seventy thousand years ago climatic conditions in southern Africa became very marginal for hominids, and there is no way of knowing whether the early symbolic expressions found along the south coast were directly linked to later developments further north or whether the nascent tradition simply died out, to be reinvented later elsewhere, even possibly multiple times. What does seem well established, however, is that the distinctive human anatomy came into existence significantly *before* we have any convincing evidence for the exercise of symbolic mental processes. Of course, there's nothing odd in that. As we've seen, the biological entity *Homo sapiens* most plausibly emerged in a fairly short-term event, one that was almost certainly not restricted to the cranial and skeletal modifications that we see in the fossil record. Quite plausibly, the genetic and developmental change that gave rise to modern bony anatomy also had soft-tissue ramifications that included some neural innovation that created a potential for symbolic thought—a potential "discovered" by its possessors only some tens of thousands of years later. In evolutionary terms such "exaptation" would have been nothing remarkable; the ancestors of birds, for example, possessed feathers millions of years before using them as a crucial element of flight, and early tetrapods acquired rudimentary "legs" in the seas long before recruiting them for walking around on dry land. Those early Levantine anatomical moderns who behaved like Neanderthals had simply not "discovered" their latent new capacity.

Since the biology was already there, the trigger for this discovery must have been a cultural one, and exactly what that trigger was is open to debate. Some people favor factors relating to the intense sociality and cooperativeness of Homo sapiens today. Others lean toward the development of the apparently unique human ability to envisage multiple "levels of intentionality" in interpreting the actions of others (for example, the capacity for me to realize that he believes that she assumes that yet another party is intending to do something). The leading candidate for the role of cultural stimulus to symbolism must, however, be the invention of language—surely the quintessential symbolic mental activity. Language depends on the creation and recombination of mental symbols, and it is hard indeed to imagine modern thought in its absence. What is less hard to envisage in the presence of the appropriate neurobiology is the devising within a small population—possibly even by kids—of structured language. An innovation of this kind could be expected to spread like wildfire, especially since cultural traits can be passed along "laterally" from one person to the next, as well as "vertically" from parent to child, like genetic ones. What makes language so unusual, and so advantageous, is that a finite number of vocal symbols can be recombined, according to a set of rules, to produce an infinite number of statements. Had the brain of Homo sapiens not been already predisposed to acquire language, that acquisition could never have been made, and, equally importantly, it is evident from bony anatomy that since its origin as a distinctive species Homo sapiens had possessed a vocal tract that was up to producing the necessary sounds. This prerequisite, too, must have been acquired in an exaptive context. Once again, for all of the evidently many temptations to view Homo sapiens as the shiningly burnished outcome of a long process of fine-tuning, we are forced to see ourselves more prosaically, as "accidental tourists" along the evolutionary trail.

Various lines of evidence combine to suggest that the behaviorally archaic Levantine Homo sapiens fossils represent a "failed" pre-symbolic foray out of the parent continent, an experiment that eventually resulted in local extinction. Clearly, the behavioral proclivity to process mental information symbolically was acquired much later, in a small local *Homo sapiens* population that had remained in Africa (possibly, but far from certainly, in South Africa). During the time in question the continent was undergoing considerable environmental stress, with widespread drought conditions. As a result of these the human population was very likely fragmented into a few favored "refuge" areas, and many human groups are likely to have experienced local extinction. As we will see in a moment, the symbolic population that arose amid all that stress may have been very small indeed; and while smallness was quite possibly a prerequisite for such a momentous innovation, it may also be something of a miracle that our ancestral population survived at all. Subsequently, in less stressful times, or as a result of its enhanced capacities, fully symbolic *Homo sapiens* expanded to all parts of the continent and, in a remarkably short time, throughout the Old World and eventually beyond. This expansion was almost certainly not exploratory in nature but was instead the passive result of aggregate population growth, interspersed with multiple contractions and local extinctions. The pulse out of Africa that ended up in Europe, in the shape of those Cro-Magnons who left us the most magnificent known record of the early flowering of the human spirit, was but one of many.

TAKING OVER THE WORLD

A scattering of fossils documents the early spread of cognitively modern humans within and beyond Africa. However, as we know it today, the bony record does not encompass enough fine detail to do much more than confirm that early *Homo sapiens* were actively spreading in the millennia following about fifty thousand years ago. If we want to look in finer grain and to inquire, for example, whether what we recognize today as the "races" of *Homo sapiens* have a definable time depth, we immediately run into trouble. For one thing, as we have already argued, "racial" features are ephemeral, and we certainly can't assume that the average morphologies typical of regional groups today also characterized their ancient counterparts of several tens of thousands of years ago. All intraspecific variants are by their nature transitory, and we can't use the same criteria used by forensic investigators today to determine the "race" of ancient human remains. An attempt to do this needlessly and hugely complicated the brouhaha several years ago over the disposition of the nine-thousand-year-old "Caucasian" fossil from Kennewick in Oregon. Wherever the proper place was to decide the affinities of this individual, it certainly wasn't in the courts.

Even in the modern world the forensic enterprise may be a hazardous one, and although anyone with a modicum of knowledge can usually make a reasonable guess at whether the possessor of a skull dug up in the woods was "black" or "white" or "Asian," quite often they will be stumped, or even plain wrong. Just as you would be much of the time if you happened to be out on the street in New York City guessing the "origins" of passersby. Particularly since it is superimposed on the variation that is naturally present within populations (some people are always going to be more "typical" of their group than others, in whatever features you might care to mention), accelerating reintegration is making it ever harder to draw those lines.

Another shortcoming of the physical record is that in many cases the archaeological evidence is insufficient to directly corroborate the notion that the spreading hominids were symbolic—the amazing Cro-Magnon achievement is a dazzling exception, though not the only one. However, one very strong hint may lie in the sheer speed of the dispersal. Modern people took over the world in an amazingly short time, displacing at least three and probably more archaic kinds of hominid as they went. This strongly suggests high rates of population expansion, and this in turn implies the sort of intensification of environmental exploitation that only the symbolic capacity is capable of bringing with it. One reason for the disappointing poverty of the fossil and archaeological records may be that the early symbolic human populations tended to spread along coastlines, and the period between about sixty and ten thousand years ago was one of intensifying cold. The huge polar icecaps "locked up" ocean water, causing sea levels to fall and coastlines to expand. Then things changed. Over the past ten thousand years or so, the establishment of warming conditions has led to the melting and contraction of the icecaps, hence to rising sea levels that drowned any sites that those who followed the coastlines might have left behind them.

Still, irrespective of the exact reasons for the less-than-ideal physical evidence for human spread, one thing is plain. The best record we have of this dramatic story comes not from fossils or material objects but from molecules and words. The DNA of every one of us carries a record of our biological past; and languages, too, contain a huge amount of information about who came from where. The ways in which languages may change make constructing trees of linguistic descent a very tricky business indeed, but comparing the DNA of samples of individuals taken from populations around the world has made it possible to reconstruct in amazing detail how human beings took over the globe. We next briefly recount this story.

WHAT THE MOLECULES CAN TELL US . . . AND WHAT THEY CAN'T

How do we extract this great story of the spread of *Homo sapiens* across the planet using molecules? We can try doing this with fossils, but due to a pretty spotty record we are left with far less than the full story. To fill in the picture we have to turn to living people and specifically to that part of living people—their genomes—that most clearly carries the history of their divergence. The good news is that living people can furnish us with incredibly large sample sizes—after all, there are well over six billion of us. The bad news is that, as we discussed in chapter 2, by doing this we are not tracking the history of individuals but rather the histories of the markers we use (we will return to this problem later in this chapter). In addition, we are limited to using what history has left us—namely, the sequences that have made it to the present—in reconstructing past events. What's more, in order to make the interpretation of existing DNA sequences meaningful, right at the beginning we have to designate that this or that sequence comes from this or that geographic locality. And while this sounds simple, it really isn't. When people living in specific regions of the world have been particularly sedentary, this approach works quite well. But where there has been substantial historic movement of people, the approach falls apart. About the only thing that *is* crystal clear is that the movement and admixture of people have hugely blurred population boundaries.

Fortunately, though, the two markers most commonly used to decipher the movement of *Homo sapiens* across the globe are the twin clonal markers: mtDNA and Y chromosomal genes. Sequences from these genes allow us to determine the patterns of maternal or female movement (mtDNA) and of paternal or male movement (Y chromosomes). But remember: these two markers comprise only a very small percentage of your whole genome, probably no more than 2 percent. So what about the other 98 percent of the DNA in your genome residing in your other chromosomes? Well, this part of your genome is much more controversial and has less utility in tracing the genealogy of humans, because it comprises the diploid part of your genome. "Diploid" simply means that this part of your genome contains two copies of each DNA sequence: one from your mother and one from your father. The reason that it is less useful than the clonal markers in revealing ancestry is that during reproduction this part of the genome not only pairs but recombines and may therefore get "scrambled" from one generation to the next. In all sexually reproducing species the complex processes that go on as maternal and paternal genotypes are combined disrupt the simple vertical transmission of the historical information encoded in the DNA sequences. And this, of course, makes population and individual histories much harder to unravel. This "autosomal" part of the human genome has nonetheless become the focus of much research concerning medicine and the treatment of disease. But we need to proceed with caution here. Predicting the probability of disease using the autosomal genome often uses evolutionary notions of descent and relies on the a priori grouping of people based on assumed histories. We will address both contexts of historical analysis in people—tracing our maternal and paternal origins and using ethnicity to deal with disease—in the last chapter of this book. But as we do so, we will need to keep firmly in mind that both aims share the same basic problem that, if not recognized and properly understood, can cause major misinterpretation.

One last point about molecules: they *can* be obtained from fossils, at least under some circumstances. And what they can tell us about extinct species is somewhat relevant to our discussion of human races. Several *Homo* fossils have been examined using molecular techniques. Perhaps the most interesting comes from the lab of Svante Pääbo and his colleagues at the Max Planck Institute for Evolutionary Anthropology and the Neandertal Genome Analysis Consortium, who generated a first draft of "the Neanderthal genome" in early 2010. Like the first draft of the *Homo sapiens* genome, this genome was generated from a combined mixture of multiple individuals. Three bones were used from the Neanderthal site at Vindija Cave in Croatia. These bones date from thirty-eight to forty-five thousand years ago, and small amounts of bone powder were obtained from each under highly sterile conditions in Pääbo's lab at Leipzig. Because anywhere from 95 to 99 percent of the DNA in such samples is non-Neanderthal, even when the samples are isolated under highly sterile conditions, techniques were developed to enrich them for Neanderthal DNA. Such techniques increase the Neanderthal DNA content some four- to six-fold. The next step was to rule out contamination of the Neanderthal DNA by DNA from the laboratory workers and other humans who have come in contact with the bones. Doing this revealed that less than 1 percent of the sequences they obtained were contaminated with modern *Homo sapiens* DNA. Next-generation sequencing technology was then used to obtain millions of small fragments of DNA sequence from the mixed samples that could be rapidly sequenced and then reassembled by computer matching. A technological tour de force if ever there was one!

What kinds of questions can we ask of these data that are relevant to *Homo sapiens* and race? It turns out that Neanderthal sequences give us extra perspective on some important questions both about reintegration and about the significance of the degree of variation we see amongst human lineages. The question of whether Neanderthals interbred with humans seems to preoccupy the imaginations of many, including scientists. The announcement of the first draft of the Neanderthal genome probably got most of its airplay because of the claimed ability of the study to address this question. Pääbo and colleagues make two important suggestions about the interbreeding question. First they suggest that Neanderthals are more closely related to non-Africans (but not specifically Europeans) than to Africans. Second, they suggest that as much as 1 to 4 percent of non-African genomes is Neanderthal in origin. Researchers who disagree with this suggestion point out that the common ancestor of the Neanderthal lineage with the *Homo sapiens* lineage existed well over five hundred thousand years ago. And because the ancestral population of *Homo sapiens* was polymorphic and contained lots of genes that *Homo sapiens* shared with Neanderthals, then by chance some genes from Neanderthals should be expected to have made it into some *Homo sapiens* genomes. This process is called lineage sorting, and it would indeed make some *Homo sapiens* individuals look more closely related to Neanderthals. This latter scenario would result in some *Homo sapiens* individuals having what you might call "Neanderthal" genes scattered throughout their genome. Pääbo and colleagues argue, on the other hand, that their observations point to a nonrandom distribution of Neanderthal genes in European genomes and that this indicates a very different process of transmission of Neanderthal genes to *Homo sapiens* than the lineage sorting process. This is why they settled on (a very low level of) interbreeding as an explanation. Determining whether or not this is the most accurate interpretation of the sequence information will require more work, but, if accurate, it would demonstrate the principle of reintegration quite nicely.

FORKS IN THE ROAD

When reconstructing history using the mtDNA and the Y chromosome clonal markers, we need above all to remember two aspects of these molecules. The first thing to remember has to do with reconstructing the history of maternal and paternal lineages in terms of a phylogenetic branching diagram. The second involves the use of an

assumed steady "tick tock" of mutational change in molecules as a sort of "clock," allowing us to place time estimates on divergences. Here again, we have both good news
and bad. The good news is that we can often obtain both a divergence pattern and a
timeframe; the bad news is that the timeframe can conflict with the fossil evidence
and that the divergence pattern needs to be interpreted very carefully. Still, if all goes
well, by looking at the branching patterns of a phylogenetic tree we can understand the
order of events that gave rise to the organisms at the tips of the tree.

Living, breathing trees need roots, and so do phylogenetic trees. A root in a
phylogeny or a genealogy tells us in which direction we need to follow the relationships that are found in a tree. Knowing where the root is tells us which branches of
the tree diverged first, then second, then third, and so on. Once rooted, the rest of
the tree gives us a roadmap, much as languages do, of the history of the entities lying
at the branch tips of the tree. But compared to the road atlas in your car, this kind of
roadmap has some unusual qualities. For a start, you can only travel in one direction
(up the tree); there are only forks (bifurcations) in the road; and the tips of the trees
are all dead ends.

In general, we do not use individual organisms as endpoints in phylogenetic trees
like this. Unquestionably, it is species, or higher taxa like genera and families and so on,
that are best represented as the tips in branching trees of this kind. Still, as we discussed
earlier, because mtDNA and Y chromosomal DNA are clonal we can validly use these
molecules to make trees of their own descent and by extension of the populations of
which they are characteristic. One advantage is that the clonal inheritance of mtDNA
and Y chromosomes decreases the effective number of copies of these molecules relative to the other molecules in the genome, and because they occur as clones in cells,
they have half the number of effective copies that other genes have. Remember that
half of nearly all of your genome comes from your mother and half from your father.
But mtDNA comes directly from your mother, and, if you are a male, your Y chromosome comes straight from your father. These two markers are therefore called "haploid"
markers, and any variant of such markers is called a "haplotype." As a result, any Y
haplotype in a Y chromosomal tree can reside at a tip on that tree, and, likewise, any
mtDNA haplotype can reside at a tip on a mtDNA tree.

Thus, in our "roadmap" we can justifiably envision mtDNA haplotypes in the
form of a tree. Now, if we know the geographic origin of these haplotypes, a hypothesis
of how the geographic variants of these mtDNA haplotypes branched off can be read

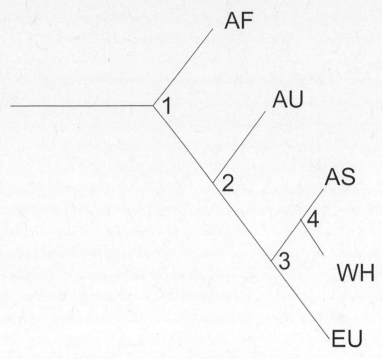

FIGURE 5. *Diagram showing the "forks in the road" for human mtDNA. Abbreviations are given in the text, and numbers at the forks refer to the numbered events in the text.*

straight off the roadmap. In the human case, as we travel along the trunk from the root, we soon encounter a fork in the road. This fork leads to mtDNA haplotypes associated with some (but not all) living African haplotypes going upward, while another fork leads downward. In pretty close agreement with the fossil record, this basic forking strongly suggests that our species originated in Africa. At the fork that goes down, we see a road leading to living Australian aboriginal haplotypes and yet another fork in the road downward. The next fork has two roads leading off it. One goes upward, and leads to some of the living Asian haplotypes. The road down leads to yet another fork. This latest fork has a side road going downward that leads to the majority of European haplotypes. But we aren't finished yet, as we still need to determine where in this sequence of events the haplotypes of native women from the Western Hemisphere (WH) come from. Something weird happens here, as the haplotypes from this part of the world come from a branch of the road that also leads to most Asian haplotypes. Okay, now try going back to the outgroup and drawing this roadmap by going through all the forks in the road. This diagram will be helpful when we use the molecular clock to talk about the times that the forks in the road represent.

TICK TOCK, DOES THE MOLECULAR CLOCK KEEP GOOD TIME?

The biochemists Emil Zuckerkandl and Linus Pauling noticed in the 1960s that the proteins they were studying had accrued changes in their amino acid sequences roughly as a function of time. In other words, they suggested that the changes they saw in these proteins were accumulating with the regularity of clocks. This idea is very simple in concept, but it contains many assumptions that need to be met before it can be relied upon for practical purposes. First, the clock needs to have external calibration points, which are usually obtained from fossil ages. Fossils can sometimes if not always give a reasonable estimate of the absolute times of divergence of taxa (although the first known appearance of a taxon can only ever give us a minimum date for its existence), and, since they are calibrated from the fossil dates, our molecular clocks give "relative" divergence times. As a result, to rely on your clock you need at least three robust fossil dates for the entities at the tips of a tree, as well as their molecular sequences. These molecular sequences are compared to each other, and the differences are counted—although in the real world some important adjustments to the raw counts have to be made. Once you have both these sequence divergences and the fossil dates, you can calibrate the clock. And then all you need to do to determine the divergence time of two species for which you don't have a good absolute fossil date (which is most of the time) is to sequence the gene you used to calibrate the clock and then translate the sequence divergence into a divergence time. Let's use a very simple imaginary example.

We have individuals of four species: a human, a gorilla, a gibbon, and a macaque monkey. We (think we) know the fossil divergence times of these species. Let's say macaques diverged from the common stem leading to humans, gorillas, and gibbons around twenty-five million years ago. Gibbons diverged from gorillas and humans about seventeen million years ago, and gorillas diverged from humans about nine million years ago. If we sequence a mythical gene (let's call it the "perfect time" gene) and we find the following sequence differences, we can then use the information to determine the divergence time of chimpanzees and humans without recourse to finding a fossil to get the absolute time. If the sequence differences between human and macaque amount to 25 percent, the human to gibbon sequence amounts to 17 percent, and the human to gorilla comparison is 9 percent, then we have a neat 1:1 relationship. That is, for every 1 percent sequence difference, 1 million years appears to have passed using the sequences of this "perfect time" gene. Now, if we are really interested in the divergence time of human and chimp and we have no fossil evidence for the divergence, we can simply sequence the perfect time gene for chimpanzee and compare it to the human sequence. Lo and behold, we find that there is a 6 percent difference in the two sequences when they are compared. Using the 1:1 calibration of the "perfect time" gene we can interpolate the human-chimp divergence as about six million years. Of course we have grossly oversimplified the molecular clock by using our mythical "perfect time" gene, but we hope you get the picture. For the most part, over long periods of time, and when you have large amounts of data, genes do seem to tick somewhat like clocks. In which case, they can be used to give us pretty good hypotheses of relative time of divergence events. But there are problems.

One of the problems with the molecular clock arises because evolution doesn't necessarily proceed in a regular clocklike fashion, as the clock model assumes. Remember the work of Eldredge and Gould on punctuated equilibrium. Here is a situation where different lineages evolve at irregular rates. In addition, not all molecules in the same lineage evolve at the same rate. For instance, the genes that code for proteins that are involved in the structure of the ribosome have apparently evolved very slowly. This slow rate of evolution happened because the ribosome is a highly conserved cellular structure that is needed to translate the hereditary material of the cell into proteins. If this cellular machine is altered even slightly, it causes widespread havoc in the translation of genes in the cell. Havoc in translation means havoc in the cell, so the proteins that make up this translating apparatus are highly conserved and change only rarely over evolutionary time. On the other hand, proteins that are involved in the functioning of our immune system are highly variable. They need to be, in order to cope with quickly arising pathogenic and other challenges from the environment. If they could not evolve rapidly, then we would have an inadequate immune response every time we were challenged by a pathogen. This is why immune system genes are among the fastest-evolving proteins in the human genome, while ribosomal proteins are among the slowest. It is easy to see that we wouldn't want to use ribosomal proteins to calibrate a molecular clock for the divergence times of higher primates. Likewise, we wouldn't want to use a rapidly evolving gene to calibrate a clock for the divergence of the major groups of Bacteria, Archaea, and Eukaryota we discussed in chapter 2.

The other really important shortcoming of molecular clocks is the shortage of really good fossil dates for calibrating absolute times of divergence. One particularly acute problem is that we often want to get a divergence time for an event that is "outside" the calibration range of the clock we are using. For instance, let's look at the "perfect time" gene again. Let's say we want to get a divergence time for the human-lemur branching. Lemurs have a much older common ancestor with us than even the oldest primate species in the "perfect time" clock we were just discussing. This puts the desired time estimate outside the range of our calibration, and tells us that we need to interpret with caution any divergence time we might estimate for lemur-human divergence. Similarly, we wouldn't really want to calibrate the divergence of human geographic groups using the "perfect time" clock that we calibrated in our example because these comparisons, too, would be out of the range of the absolute time fossil calibrations, at the other end of the scale.

Problems like these can often be overcome by correcting either for events that make the clock for a particular molecule irregular or for estimates outside the calibration range. We have, in fact, already mentioned one approach to doing this. Remember that two entities at the tip of a tree are usually viewed as descending *from* a common ancestor. But as we mentioned in chapter 2 they can also be viewed as coalescing *to* a common ancestor. When you use this latter trick, you can take into account many of the demographic aspects of the entities you are looking at, such as migration rate, population size, and other population genetic parameters. This will help give more accurate estimates even if the clock is not ticking regularly, and we now have pretty robust estimates for the divergence times marking many of the major events as human beings spread throughout the world.

THE HUMAN JOURNEY

There are several ways to view the vast amount of information that has been and is being collected on the genetics of human movement and migration. The first way is to simply look at what the patterns tell us about how we moved about the planet. The second way is to look specifically at the variation inherent in our populations to interpret process. These two ways of looking at genes give us different inferences about our history. The pattern way of interpreting things has led to a very clear and well-accepted picture of our movements in close agreement with the paleontological evidence. Specifically, we have seen that the human tree is rooted in Africa, and we can readily say that the first migration out of Africa of mtDNA haplotypes went all the way to Australia and occurred about fifty thousand years ago. Not too much later, a second migration out of Africa occurred that resulted in colonization of Eastern Asia and Asia Minor by *Homo sapiens* at around forty thousand years ago (at which time the last record of the long-resident *Homo erectus* occurs). This migration into Asia was followed by migration of mtDNA haplotypes from Asia into Europe at about thirty thousand years ago. Interestingly, this is about the same time that our extinct relative *Homo neanderthalensis* disappeared. There was then a relatively long period before *Homo sapiens* crossed the Bering land bridge, about fourteen thousand to seventeen thousand years ago, and colonized the Western Hemisphere. The haplotypes of Native Americans make an interesting subject in themselves, appearing for the most part to arise from a divergence event in central Asia. Last but not least, there is the concerted colonization of the Pacific at about four thousand years ago by mtDNA haplotypes that are closely related to Asian haplotypes. Not all of these times come from molecular clocks, because the fossil and archaeological records also allow for strong absolute timing of these migrations.

So far we have only mentioned the female lineages, but at the gross level the male lineages did pretty much the same thing. Yes, this might simply mean, of course, that the men were following the women. But while it is similar, the pattern of the Y chromosomal tree is not identical to the mtDNA one, indicating a somewhat different evolutionary history for the paternal lineages compared to the maternal ones. We will return to this in a moment, but first let's look a bit more closely at what we can say from the molecules about human migrations around the globe.

Perhaps it is best to start this inquiry by asking an obvious question: What do these patterns of relationship and migration have to do with our genetic conceptualization of race? Logically, one might expect that the major past migration events would map directly on to the way we humans have since characterized races. The older populations stayed in Africa, with one great migration to Asia (the subsequent migrations from Asia, one to America and one later to the Pacific, can be considered offshoots of the Asian migration) and another to Europe. And we'd have three major races, end of story. But remember, we're dealing with human beings here, and *Homo sapiens* is nothing if not capricious. So, as it turns out, things weren't at all that way. And this is where the second way of looking at the vast information on human diversity comes in.

From 2007 to 2009, over one hundred papers on human migration and population structure were published in prominent scientific journals such as the *Annals of Human Biology*, the *Journal of Human Genetics*, and the *American Journal of Physical Anthropology*. These articles typically constitute what you might call "micro-surveys" of human movement and ancestry, addressing very specific groups of people in an effort to characterize their mitochondrial DNA and Y chromosomal makeup and to add to the number of members of our species who have been surveyed in this fashion. Usually, such studies focus on a specific ethnic group the researchers are interested in, for example the Malagasy people in Madagascar, the Qinghai people of Tibet, the Han people of China, the Druze from the Near East, or Polish and Russian folk from Europe. Studies have included both highly admixed populations such as some from Mexico, and not-so-admixed ones, like that of El Salvador. Once characterized, the group of interest is then situated within the larger context provided by the amazingly detailed mitochondrial and Y chromosomal databases accumulated in surveys worldwide over the last couple of decades. Sometimes the research goal is simply to characterize the makeup of these populations for forensic purposes—a practice we feel is misguided and to which we will return in our concluding chapter. More often, though, the aim is to understand the dynamics of how these populations have interacted, and continue to interact, with other populations. Most of the time a genealogical tree is constructed, in order to demonstrate the relatedness of the mtDNA and Y chromosome DNA of the ethnic group under study to those of other groups that are geographically or historically associated with them. Other approaches use the coalescent method we described in chapter 2 for analyzing the migration and admixture of populations.

Whatever the methodology, these studies rely on certain assumptions. One of these is that the ethnic groups of interest are cohesive enough to be studied using current methods. Even if the researcher knows that the population or ethnic group of interest is admixed, he or she still needs to assume that the ancestral populations that gave rise to the admixed population were cohesive. Such assumptions of cohesion are large indeed. Yet without some level of genetic cohesion, studies like these are not possible, purely because the algorithms used to analyze the data must assume that it exists. What's more, a strong database of known haplotypes for mtDNA and Y chromosomes is also needed to conduct these analyses properly. So another big assumption is that the established database reflects some aspect of reality. Finally, and most alarming to us, is the assumption that the patterns described by the mtDNA and the Y chromosomes accurately represent the patterns of *people*. Not all studies of human ethnic groups make this assumption, but it is nonetheless very easy to fall into the trap of making claims about entire peoples who share an ethnic identity simply from haploid clonal markers that represent only a fraction of their entire genomes. If you are not very careful you can end up making broad claims about historical processes when all you are actually doing is characterizing maternal and paternal markers.

This having been said, it remains true that there is invariably a certain amount of cohesion of the ethnic groups being studied. As a result, much of the information that has been collected contains a genuine historical signal and is thus extremely useful in demonstrating general patterns in the expansion of our species across the planet. What's more, a look at the literature published in the past two years or so shows a pretty even distribution of information published on peoples from all over the world. And almost all of these studies delving into the processes behind our distribution across the globe came to a uniform conclusion. Which is, quite unsurprisingly, that we adore having sex with each other, no matter what we look like. In all of these papers, we find the key words *admixture* and *expansion* used over and over again. In other words, no matter how much *Homo sapiens* explores and moves about, we like to mate with whatever other people we meet up with. And with this reality firmly in mind, let's take a continent-by-continent stroll across the globe, to see how these studies have impacted our understanding of the genetics of our species.

JOURNEYS AND POPULATIONS

Our first stop is in Africa. In human terms, this great continent is a hugely diverse place. There are three major genetic observations that have been made about the diversity of people living on the African continent. First, Africa shows more genetic diversity than the other continents. Second, most of the genetic variation outside of Africa is a subset of the variation found within Africa. Finally, genetic diversity decreases with increasing distance from Africa. Africa is also diverse in other respects. With nearly 1 billion people living on the continent, there are over two thousand linguistic groups. What's more, it isn't rare for African individuals to know four or five different languages, some from completely different linguistic groupings.

Studies of diversity in Africa have revealed that three major mitochondrial DNA lineages evolved in the continent concurrently with the major mtDNA lineages that made the excursion out of Africa. These lineages are old, to say the least, in comparison to those that migrated out of Africa. Recently, Sarah Tishkoff at the University of Pennsylvania, and her colleagues examined over 1,300 nuclear genetic markers for about two hundred people. The majority of the subjects were from Africa or were of African descent. The researchers aimed to analyze the nuclear genomes of representatives belonging to as many linguistic groups as possible, and they used many non-Africans as comparisons. This was a multifaceted study, but in our view its main value lies in its analysis of genetic variation among African people. Among other things, the variation patterns reveal that nuclear genes from people of Niger-Kordofanian ancestry are widespread, meaning that the agriculturally inclined Bantu-speaking groups have had a large impact genetically on African population variation and structure across the continent. The dispersal of these genetic markers across Africa started approximately five thousand years ago, and it has continued to the present day. Yet more significantly, the study suggests that even though some groups (such as the Maasai and the Pygmies) maintain strong cultural cohesion, they nonetheless show large amounts of genetic variation as a result of population mixing. The bottom line here is that culturally individualized African populations may be incredibly complex, much more so than we had previously thought. We clearly need to resist the tendency to oversimplify the patterns that exist in populations, and the apparent extreme genetic complexity even of culturally homogeneous African populations is a great example of why that is.

An equally important aspect of understanding "African" diversity involves the many studies aimed at assessing the genetic consequences of the terrible institution of slavery. These studies indicate that the genetic structure of African American populations is extremely different from that of the ancestral African populations, a result that is hardly surprising, given the diaspora and subsequent intermixing, in new environments in the Americas, of African people stolen from their homelands. It is more surprising, perhaps, to read of the finding that "fewer than 10% of African-American mtDNAs matched mtDNA sequences from a single African region suggest[ing] that few African Americans might be able to trace their mtDNA lineages to a particular region of Africa." Yet another sad consequence of the African diaspora.

Another example of the admixture of people in the African region concerns the peopling of Madagascar, the second-most-recent large landmass to be colonized by humans (under two thousand years ago; the most recent such colonization was that of New Zealand, under one thousand years ago). It has long been known that the modern Malagasy population, which is quite culturally and linguistically homogeneous, is nonetheless the result of admixture of people from two hugely geographically distinct regions: sub-Saharan Africa and Southeast Asia. Now that DNA studies are available, the extraordinary extent of the resulting genetic diversity is becoming even more evident; indeed, the more Malagasy people that are included in these studies, the more complex the inferences concerning the origins of the people living on Madagascar become. Apparently, the degree of human mtDNA variation on this large and glorious island is nearly endless.

Our next stop is Asia. On current information there are as many as four major migrations or occupations that we need to keep in mind when discussing this vast region and its island appendages. The first is the migration along the Indian subcontinent and eventually into Australia. The second is the foray that peopled eastern Asia (and subsequently the Pacific). The third is the migration to Siberia and the subsequent movement of people to the Americas. And the fourth concerns the processes affecting the genetics of people in Asia Minor and the Arabian Peninsula. The ethnic groups in all of these regions are numerous and complex, and phenotypically diverse. But again there is a common theme, and that is *admixture*. Let's look briefly at each of these demographic phenomena individually.

First, Australia. The evidence for the idea that the ancestors of today's Australian aborigines arrived in Australia via Africa and India comes from the sharing of a specific mtDNA lineage between peninsular Indians and native Australians. The mtDNA lineage is called M42, and before studies that were published in 2009 it was only found in Australian aborigines. The corridor for this migration was most likely provided by the so-called "southern route" along the shores of the Indian subcontinent. Today, though, Indian populations are highly variable. One population that has attracted a lot of attention lately is the Muslim population from northern India. Because Islam arrived in India from the west, researchers expected to find the signature of western Asian genes in these people. However, the research done so far indicates that there is a conspicuous lack of any western Asian gene contribution to the Muslim population of India. Indeed, the researchers inferred that the spread of Islam in this area was a "predominantly cultural transformation associated with minor gene-flow from West Asia."

Next, eastern Asia—and ultimately the Pacific. A lot of information now exists relative to the genetics of people from East Asia. Frequently, population genetic analyses are confounded by admixture, but the general pattern of initial migration to the north and west and the subsequent filling of East Asia by migration from north to south and from west to east are quite evident. Most recent studies have aimed at deciphering the origin of different Chinese ethnic groups, with an alarming (to us) increase in studies for forensic identification of people from the various ethnic groups that exist in China. As we have discussed earlier in this chapter, the use of mtDNA to identify people with a population of origin is often misguided. Studies without this specific purpose tend to point toward extensive interbreeding of people in this part of Asia. The Daic and Hmong-Mien peoples, for example, have been characterized as having "frequent intermixing taking place throughout." In another study the Pinghua people of Guangxi in China were examined for relatedness to other Chinese ethnic groups. The results indicated that, genetically, they had been thoroughly "assimilated by the Han" through admixture. The Tu people, one of the many other minority groups in China, were also studied, with the result that they appear to be "multi-origin and also merged with other local populations." An examination of the population genetics of people along the Gansu Corridor, which is considered the easiest western access to the Far East, via the Silk Road, revealed that "both European-specific haplogroups and Eastern Asian-specific haplogroups exist in the Gansu Corridor populations." And moving a little south, the populations of eastern Indonesia have been characterized genetically as having gone through "several admixture episodes." DNA studies make it pretty clear that the peopling of the Pacific by today's Polynesians occurred as a result of a migration from East Asia, but in accordance with the pattern established by other humans migrating to new areas, "Polynesian ancestors originated from East Asia but genetically mixed with Melanesians before colonizing the Pacific." Another story of admixture shaping the genetic makeup of a major movement of humans.

Northern Asia is home to over twenty major linguistic groups and is considered quite complex ethnically. We include Mongolia in this region, as a starting point for examining the genetic processes involved in the peopling of this part of the world. A large number of native Mongolian people have been examined for mtDNA, and, not surprisingly because these people are traditionally nomadic, the "data suggest that the Mongolian empire played an important role in the maternal genetic admixture across Mongolians and even Central Asian populations." The Yakuts from the northeast of

Siberia display a strange difference in their mtDNA and Y chromosomal diversities. The discrepancy in diversity is the result of different evolutionary histories of these two molecular markers that are exactly the same for what we describe in chapter 5 for Charles Darwin's pedigree. The genetic diversity of the Tatars in the Tobol-Irtysh basin indicates that, as with other Tatars, they are highly genetically diverse as a result of "various interethnic relationships and different ethnic components integrated into these groups." The Tatars from this region appear to be much more variable than Tatars from the other two geographic regions where these people live. In another study of people from the Nganasan, Yukaghir, Chuvantsi, and Chukchi groups, and of Siberian Eskimos plus Commander Aleuts, the admixture of these peoples was also plainly evident.

It is from the Siberian ethnic groups that people next migrated to the North and South American continents, leading to the rapid peopling of these twin vast areas. Population genetic studies of people from both continents reveal a high degree of admixture, not only amongst the first migrants to those continents, but as a result of the later intrusion of Europeans (our last stop) into the western hemisphere. In certain enclaves this has not turned out to be the case: a mtDNA study of native Americans from El Salvador revealed that there has been limited admixture of people from this region of the New World with African or European women. This local pattern nonetheless contrasts with the story in many other areas of the New World: for example, a study of native Mexican people revealed that although they "show a closer relationship to native North American populations, they cannot be related to a single geographical region within the continent." This of course indicates large degrees of admixture.

Far away to the south and the east in Asia, much work has been reported in the last two years on the population genetics of people living in the Arabian Peninsula and Asia Minor. Perhaps nowhere else in the world is there more ethnic awareness than in this area, so the population genetics of the region is of considerable interest. Hundreds of individuals have been analyzed genetically, and some very interesting (though not necessarily so surprising) results have been obtained with respect to admixture. For instance, people from Saudi Arabia show high affinity to Eurasian ethnic groups, an observation explained by continued expansion and admixture of people across the peninsula, as a result of its "being more a receptor of human migrations." Today's Yemeni people, while fairly distinct in their population genetics, have also "received West Eurasian, Northeast African, and South Asian gene-flow."

To the north and east, the Altaian Kazakhs of southern Siberia have also been genetically characterized, with results suggesting the existence of "complex interactions of the Kazakhs with other Turkic groups, Mongolians, and indigenous Altaians" via admixture and expansion. For the ethnic Armenians in this area there is "more structure for the Y chromosome than for the mtDNA," again harkening back to Charles Darwin's pedigree and implying that the two markers have experienced quite different evolutionary histories. One of the things that these very localized genetic studies cannot do is to assess the degree of variation existing between differentiated populations. This is because these studies are focused on the population genetics of very distinct ethnic groups, and cross-comparison is not possible. But in cases where cross-comparison *is* possible, such as in a population genetic study of several ethnic groups in Central Asia, the "Lewontin effect" is almost always observed, reflecting the observation we have already mentioned by Richard Lewontin that "there are more differences between populations of the same ethnic group than between ethnic groups for the Y chromosome."

Europe is a smaller and less geographically cohesive "continent" than Asia, but it, too, has a unique genetic history. Modern humans began populating this continent, from both Africa and Asia, about forty thousand years ago. Again, some of the available analyses of populations from this continent have been accomplished for "forensic" purposes, but many have been designed to add to the growing databases for medical research.

Due to the relative isolation of the Scandinavian countries, groups of people from Finland and Iceland have been the object of intense genetic examination in the expectation that these populations would prove to be genetically quite homogeneous and thus would be good subjects for demographic investigations of various kinds. In the case of Finns, however, it has turned out that substantial gene-flow has occurred in the southwestern part of the country, although not in eastern areas. This gene-flow appears in the Y chromosome record and thus is male-biased. But the bottom line here is that migration and admixing are both important, and unexpected, factors that the medical geneticists now have to deal with. A similar expectation of homogeneity applied in even greater degree to Iceland, making this country host to the "ideal" population for understanding medical genetics. This expectation emanated from origin myths of a single founder event, from the supposedly low degree of migration to the island, and from the meticulous genealogical records kept for the country. The prevailing belief that the Icelandic people are a cohesive ethnic group, and that their genetics would reflect this, led to the establishment of a company called deCODE. The subsequent analysis of the population genetics of Icelanders revealed an unanticipated degree of variation in people from the island. In fact, Icelanders were found to be every bit as variable as the frequently invaded French. Yet deCODE still claimed to be one of those companies able to type you to your continent of origin (a geographic way of saying "to your race").

Other studies in Europe have focused on very specific genetic questions regarding the peopling of the continent. For instance, a genetic analysis of the peopling of Sicily reveals the "presence of a high number of different haplogroups in the island" that "makes its gene diversity . . . reach about 0.9." The most diverse populations possible have gene diversity at 1.0, and the proximity of the Sicilian population to this maximal gene diversity measure means that Sicilians have unquestionably had an extremely complex history of migration and admixture. Another study of populations in southern Italy indicates that people from this area, too, are highly diverse genetically, as, indeed, are many other Mediterranean populations (hardly surprising for the "crossroads of the world").

On a different time scale, another study of Italian populations demonstrates the fragility of genes in populations as a result of migration and admixture. The study in question involved a comparison between native Tuscans buried between the tenth and fifteenth centuries AD, current residents of Tuscany, and descendants of the Etruscans, the Bronze Age inhabitants of Tuscany. The results indicate that the current Tuscans show no "clear genetic relationships" to the Etruscans, the original inhabitants. This example points to the ease with which genes from populations can be chased from specific geographic areas as a result of secondary contact with other populations and subsequent genetic replacement or assimilation. Such displacement and replacement does not invariably happen, but we would hazard that it is the prevalent mode by which humans have moved about the planet.

One final recent study on the genetics of Europeans has gained a lot of press lately, particularly because of its claim that the authors could identify the ancestry of people in Europe. This study used over five hundred thousand SNPs ("single-nucleotide polymorphisms," small genetic substitutions we will discuss at length in chapter 5) on 3,192 individuals (mostly from Switzerland, France, Spain, England, and central Italy) to plot the genetic distance of individuals relative to their geographic location in Europe. The resulting graph looked a lot like Europe itself, with clusters of points for Spanish-speaking individuals residing in the lower left of the graph, English speakers on the upper right, French speakers in the middle, people who identified their origin as Switzerland below the French, and the central Italians below them. In some cases this study claims to be able to assign an individual to within four hundred kilometers of their geographical point of origin—but only with some interesting manipulations of the data. Thus, while this claim may be valid in some cases, when the data are examined more closely, assignment at a 99 percent confidence interval (and really you would want better) for most ethnic groups in the study assignment can only be accomplished to within eight to twelve hundred kilometers. In some cases, where sample sizes fluctuate, the accuracy is even worse.

The ability to type people to a particular region is a significant claim, so let's take a closer look at the study and at its implications. We suggest that this study is an excellent example of our major thesis in this book—that there is indeed sometimes a signal, but that this signal is eroding. There are two concerns here, both involving data "trimming." In the first instance of data trimming in the European study, the authors removed many of the half-million SNPs because they "showed evidence of high pairwise linkage disequilibrium as well as unique genomic regions (such as large polymorphic inversions) that might obscure genome-wide patterns of population structure." This practice of removing genetic information to "clean up" the results is common, and in some cases it can be justified because the underlying pattern of a system can indeed be obscured by messy data. However, removing data that "obscure genome-wide patterns of population structure" may also result in removing data that tell a more complete story of the genetics of the people on this continent.

To demonstrate this we show two trees constructed from a different data set in figure 6. The genetic data set is a large one (26,530 genetic markers) generated by Mark Shriver and his colleagues and analyzed by Rick Kittles and Kenneth Weiss. The individuals comprising the samples are from the three major geographical areas of the planet (Eu = Europe; Ch = China; JP = Japan; Af = Africa). The tree on the left is one in which information in the data set that shows no population structuring has been removed (only the top one thousand structured genes are used to build the tree). In contrast, the tree on the right was constructed from the one thousand "worst" genes with respect to population structure. While we don't doubt that there is structure present in the top one thousand genes, we think it equally important to doubt that there

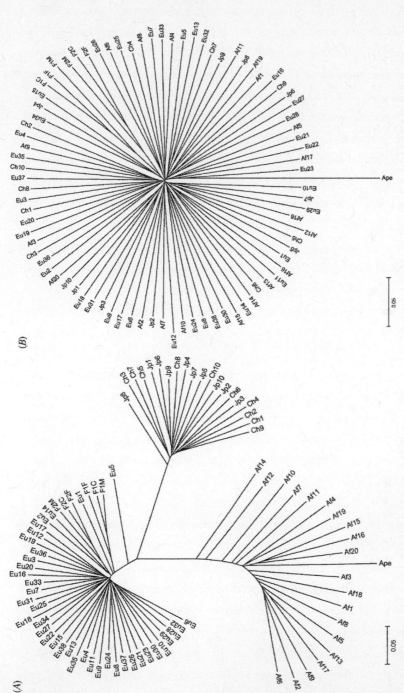

FIGURE 6. Two trees generated by similarity linkage. A) Tree generated using the top 10 percent of genes that imply structure between the three implied "racial" groups; B) Tree generated using the bottom 10 percent of genes that imply structure between the three implied "racial" groups. Each branch leads to an individual, with letter and number abbreviations. Abbreviations for individuals are A=Africa; Eu=European; Ch=Chinese; Jp=Japanese. The tree is rooted with disease/variant using an ape (chimpanzee). The figures are taken from Kittles and Weiss (2003) and were generated from data presented in a study by Shriver et al. (2003).

is no structure in the bottom one thousand genes. And how does one represent the discrepancy here? Well, if we return to our initial claim that it is invalid to draw trees of individuals that are reticulating (are reproducing with each other), then there is no problem. No tree, no problem. But there is a huge temptation to represent these data with a tree. One way to get around this is to use the graphing approach that John Novembre and colleagues accomplished on the most recent large dataset for Europeans. Still, it turns out that data trimming is critical for either approach, and the Novembre group also removed individuals from the data set that they claimed were "outliers." Table 3 shows how the authors trimmed individuals from their study and the numbers that they are left with.

TABLE 3. Sample size diminution in the Novembre et al. study

Sample size	Stage of analysis
3192	Total individuals of European descent
2933	After exclusion of individuals with origins outside of Europe
2409	After exclusion of individuals with mixed grandparental ancestry
2385	After exclusion of putative related individuals
2351	After exclusion based on preliminary run

Thus, after all of this "thinning," the researchers removed nearly 800 individuals out of 3,200. In other words, 25 percent of the individuals were rejected because they were "outliers." As if this weren't enough, another 1,000 or so individuals were removed to "reduce" the data set further for ease of computation. But we suggest that removing "outliers" like this causes you to miss something incredibly important about human populations at the genetic level.

Sure, when you "cherry pick" a data set like this, you will obtain an answer that agrees well with the rules under which you did your cherry picking. But the really interesting aspect of this study is actually the 25 percent of the population that was left out. What does an "outlier" mean here? In the real world outliers are, more than likely, migrants or admixed people. The large proportion of individuals (a quarter!) who fall into this category is very significant in and of itself, and it may in fact even be larger than indicated in this study. Many of the 1,300 or so individuals left in the study at the end of the "thinning" process could actually also be removed to make the results even cleaner. But to get the assignment of individuals to origin to be 99 percent accurate to within eight hundred kilometers, you might need to remove another 600 or so individuals as "outliers." And this, of course, would mean that as many as 50 percent of the population are in the interesting category of people who have migrated or admixed. To us, this way of looking at the variation produces a very much more interesting result and provides a much better indicator of reality. In fact, as we pointed out earlier in this chapter, any appearance of structure is fated to erode in the next few centuries (should our species make it that long), and we profoundly doubt the utility of structured analysis as a long-term way of describing human populations.

The structures of relatedness suggested by the figures above are also interesting because they allow us to address Lewontin's famous observation (of much greater within-population than between-population variability among humans) more succinctly and to make it clearer whether, as has been charged, this is a "fallacy" rather than a real observation. To us, clarifying this problem is very important since we consider Lewontin's original observation, and the failure to reject it experimentally in the last four decades, as one of the strongest arguments against the biological existence of "race." For it turns out that Lewontin's observation is a fallacy only if one has a previously defined concept of "difference" between genomes.

The generally accepted way of defining populations is to use a "genetic distance" metric, as in figure 6. However, as Witherspoon and colleagues argue, and the figure shows, a great deal of genetic similarity between individuals belonging to different clusters is obscured by this approach (compare the resolution of the tree on the left with the lack of resolution in the one on the right). This happens because when you use "multi-locus" clustering methods like those used to create the figure, the clusters are based on population-level similarity and not on individual similarity. The method of clustering then leads to the classification of individuals on the basis of their similarity to the typical gene allelic makeup for any of the populations being examined. Lewontin made his statement about using any single randomly chosen locus, and in this sense his statement is not a fallacy. As Witherspoon and colleagues point out, "The fact that, given enough genetic data, individuals can be correctly assigned to their populations of origin is compatible with the observation that most human genetic variation is found within populations, not between them. It is also compatible with our finding that, even when the most distinct populations are considered and hundreds of loci are used, individuals are frequently more similar to members of other populations than to members of their own population." Lewontin's original statement thus remains the single most relevant observation ever made about human populations at the genetic level, and it dramatically underlines the lack of biological underpinnings for the concept of race.

WHO WAS EVE?

Having looked briefly at the peregrinations of *Homo sapiens* around the world, and at the rather untidy traces those movements left behind, let's look back once more and ask what may have happened before the sixty-thousand-year divergence point when *Homo sapiens* first migrated out of Africa. At Washington University in St Louis, Alan Templeton has used the coalescent method to estimate the timing of what he calls "range expansions." These range expansions are, simply put, significant migration events in human history. Templeton used several nuclear gene haplotypes to discover that there have been three major out-of-Africa range expansions in the genus *Homo*. The first of them took place about 1.5 to 3 million years ago and involved the migration of what he calls *Homo erectus* (at any rate, an early species of *Homo*) out of Africa. The second occurred about 500,000 years ago, and involved the expansion of the Acheulean culture. Broadly, this agrees with paleontological evidence for the penetration of Europe by *Homo heidelbergensis* at about this time. Finally, at about 170,000 years ago came the most recent out-of-Africa expansion, leading Templeton to suggest that *Homo sapiens* came out of Africa repeatedly. Intriguingly, the picture has recently been augmented by the identification, in a single Siberian finger-bone some 50,000 to 30,000 years old, of a unique mtDNA signal. This suggests that yet another lineage may have come out of Africa at some time after about one million years ago. Interestingly, penecontemporaneous Neanderthal fossils, also identified through their DNA signature, are known from a site not far away. Because the bone that yielded the strange DNA is morphologically undiagnostic, it is impossible to assign this specimen to any known taxon. However, we know that not so very long before this time Europe had hosted both *Homo heidelbergensis* and early members of the Neanderthal lineage, so it would not be very surprising if this new Siberian fossil represented a later-surviving descendant of *Homo heidelbergensis*.

Another interesting aspect of the divergence of Homo sapiens at the 170,000-year-time-point is what others with media savvy have called the "mitochondrial Eve hypothesis" and what Templeton simply calls a logical outcome of coalescent theory. When Allan Wilson and his colleagues first used mtDNA to look at the genealogy of female lineages in the 1980s, the popular media picked up the story and unfortunately reported that the mtDNA "Eve" had been found in this general time range. This astonished the general public, but as Templeton emphasizes it is really just a simple product of the coalescence of several lineages back to the single common ancestral haplotype. It's merely how coalescence and evolution work. If the data hadn't shown the coalescence to a single haplotype, then that would have been news. The "discovery" of mtDNA Eve was predictably enough followed by the discovery of "Y chromosomal Adam." And in a sense, mtDNA Eve and Y chromosomal Adam are the ultimate ancestors that connect us all, though they are haplotypes and not individuals. Along the way to the present, there have subsequently been thousands, or more probably millions, of ancestral and descendent haplotypes. What's more, though you can indeed find a "most recent common ancestor" for any particular gene or DNA sequence, different genes may well give different results, owing to the number of types of event that may contribute to a particular history. In this perspective, you might expect to find multiple Eves or Adams representing various points along a historical trajectory.

It was, in any event, in the brief period since the initial emergence of our species, probably at less than 200,000 years ago, that all of the physical variety that we see today in Homo sapiens was acquired. And it was in the even briefer period following around 60,000 years ago, that all of the physical variety we encounter outside Africa came into existence. We cannot tie in particular physical features to specific clonal marker haplotypes, but if you need more convincing evidence of the superficiality of those physical differences than their extreme recency, it is hard to see what it might be.

We will look at the little that can be said about the broader biology of our differences in the next chapter. Right now, suffice it to note that, for all the differences we perceive among ourselves, we are a recently evolved and very closely knit species. Indeed, there are more diverse mtDNA haplotypes in chimpanzees just from West Africa, than there are in all of Homo sapiens.

CHAPTER 4

IS "RACE" A BIOLOGICAL PROBLEM?

HUMAN BEINGS vary. That is no surprise. Everyone knows that some people are better poker players than others, that pearl divers can hold their breath longer than bus drivers can, and that some landlubbers can shimmy up coconut trees without any apparent effort, leaving the rest of us scratching our heads at the bottom and wondering how they do it. Over the years the many ways in which people can and do vary have been intensively scrutinized, at the individual level as well as at the "racial" one. In the former case you have the problem that is always present when you study unique phenomena: a lack of perspective. We remember years ago reading a research article co-authored by at least a dozen different scientists but devoted to one single outstanding runner, Peter Snell. It was impossible not to conjure up images of Snell standing patiently like Michelangelo's *David*, while a legion of gnomes in white lab coats scurried up and down ladders braced against his torso, brandishing tape measures and calipers. The image was hilarious, but it didn't disguise the fact that all you could ultimately conclude from the scientific data presented was that Snell was a remarkable individual—and the next question was, of course, "compared to whom?" For in the end all individuals are simply floating points, and to make sense of them you have to place them within a group. Yet we've already seen—and will see again, in more detail—how frequently it is difficult or impossible to assign individuals to groups that can usefully be compared—especially groups defined by more than one trait. And if you can't characterize your groups effectively, you're pretty much at a dead end.

So while looking at variations purely among individuals is a lot easier than defining groups, on its own it is a pretty boring exercise that would probably make your cocktail party companions move discreetly but rapidly toward the bar as soon as you raised the subject. Worse, if such tedium bothers you, and you want to go any further than mere observation, things rapidly get complicated again. This is because humans are biological creatures, which means that every individual is the outcome of a long and intricate process of negotiation between its genotype and the environment that surrounds it. So as soon as you have identified a difference and want to start asking such questions as "why?" you run smack dab into the problem of knowing whether the attribute in question is acquired, or inherited, or something in between. Does it result from years of painstaking and sometimes painful practice, or is it simply the genetic luck of the draw? Or—in the event the attribute in question is a desirable one—is there anything you can do to improve your own performance, even if optimization may be out of reach?

Well, only identical twins have precisely the same genomes, and everybody, without exception, will have had different backgrounds and experiences. Even in the case of those identical twins, birth order will have made a difference, right from the beginning. As a result of complications like this, any potential intrinsic and extraneous influences on what you are observing are very difficult to control for. What's more, even if you can devise a way of controlling for them, the attributes at issue must be accurately measurable if you are going to compare them in any scientifically acceptable way. And it turns out that the range of attributes that fulfill the conditions necessary for comparisons of this kind among human beings is smaller than you might think.

One area of human experience where performance is rigorously measured in a quantitative way is sports, and two scientists at Penn State University recently did an intriguing study on human sprinters. This was a great subject for investigation, because it involved both a body system that has been of critical significance in human evolution and a complex mechanical lever arrangement in which potentially critical variables can be accurately measured. It has long been known that, compared to most of us, sprinters tend to have a larger proportion of "fast-twitch" fibers in their leg muscles. These muscle fibers contract more rapidly, but fatigue more quickly, than do "slow-twitch" fibers, which continuously burn oxygen and can keep repetitively contracting for a long time. The question that the Penn State researchers posed was whether these differences in muscle fiber proportion were matched by any anatomical disparities between sprinters and others in the foot and lower leg. Their answer was a resounding "yes!"

Compared to a control group of more sedentary types of similar stature, a sample of a dozen sprinters had longer toes and shorter shins and heel bones. The long toes were not unexpected; they provide more sustained contact with the ground as the foot extends, much like the nonretracting claws of cheetahs, which are the fastest of all mammals. More surprising were the short heel bones. Powered flexion of the foot at the ankle joint, pushing the toes down as the ball of the foot takes the weight of the body, is achieved by contraction of the calf muscles that pull on the Achilles tendon attached to the back of the heel. A longer heel would provide those muscles with better mechanical advantage on the pivot in front of it. So the expectation was that the sprinters would have long heels to add power to the toe-off. Yet the researchers actually found that the reverse applied. Why? It turned out that the movement of the Achilles tendon as the calf muscles contracted was some 25 percent shorter than in the couch potatoes (well, we don't really know that it would be fair to call the controls couch potatoes).

Fast-twitch muscle fibers contract with less force than slow-twitch ones do, and it appears that recruiting the slow-twitch fibers helps overcome the mechanical disadvantage of the short heel-lever, especially since the muscles have to be contracting for a shorter length of time to produce the same tendon movement. The fast-twitch muscle fibers help with speed, but the added power of the slow-twitch fibers is also necessary for optimum sprinting performance. The bottom line is that, in this case, it is indeed "in the genes": you may be able to improve your sprinting performance by rigorous training, but whatever "race" you might or might not belong to, you're unlikely to make gold if you don't have the requisite anatomy. Just as you'll never be a world-class accountant if you don't have a good head for figures or a prima donna if you're tone deaf. Competence is something that can usually be striven for; excellence is a gift from the gods, even if it may not come easy.

Now, all athletics aficionados know that, currently, there is a curiously large proportion of world-class sprinters who are at least ultimately of West African origin, while eastern Africa has produced what seems like a disproportionate number of outstanding endurance runners. Should we, then, expect that West Africans as a whole should tend to have longer toes and shorter lower legs than East Africans (or the rest of us)? And do East Africans, who excel at a form of running that makes demands on the body that are very different from those of sprinting, have some comparable physical tendency that explains why they are the world's standout long-distance specialists? The question intrigues, but the response has to be that we simply don't know; and anyway the answer would almost certainly be poorly predictive about any particular individual.

Would it have helped clarify the situation if the authors of the Penn State study had identified their subjects to "race"? Almost certainly not, because if any of them do indeed have African antecedents, in the United States they would certainly have been classified as "black," confounding at least two groups with stellar but entirely different athletic records. What's more, each group already has blurry edges that are only becoming more so. The moral here is that what you find may well often turn out to be in part or even entirely an artifact of how well you can define the group you are looking at. "Sprinters," though limited, is pretty good, because it involves both a clearly definable and accurately measurable ability and because it does not involve any *a priori* classification of its possessors that is unrelated to that trait. "Blacks," on the other hand, or even "sub-Saharan Africans," would fail on both counts. And in any case, it's impossible to know to what extent transient cultural influences are involved in the patterns we see today. Maybe it will all be different tomorrow, even without the mixing that is going on.

There's another classic example of apparent adaptation that clearly transgresses "race" boundaries. Human beings are mammals, suckling their young. The major sugar in mothers' milk is lactose, which must be digested by the enzyme lactase. Lactase is produced by all mammal infants, but in most cases its production wanes or ceases after weaning. After all, beyond that point most mammals are never again exposed to lactose in their diet. Most human beings are perfectly normal mammals in this regard, and as a result have problems digesting milk as adults; yet in some human populations, many or most individuals continue to produce lactase into adulthood and indeed throughout life. Why? It turns out that extended production of lactase is pretty highly correlated with cultural practices that involve cattle-rearing. Lactose tolerance (also known as lactase persistence) is common in human populations that consume a lot of milk. Among northern European populations, only about 1 to 15 percent of individuals are lactose intolerant, and the lowest frequencies are found in populations that depend heavily on dairy products—the Basques (0.3 percent) and the Dutch (1 percent), for example. Interestingly, analyses have shown that the genotype producing lactase persistence in northern Europeans is very different from its ancestral human equivalent. Outside Europe, the lowest frequencies of lactose intolerance are found among African cattle herders (around 20 percent), whereas in other African groups levels it can reach almost 90 percent. Non-milk-drinking Native Americans reportedly come in at nearly 100 percent, closely followed by the dairy-phobic Chinese, at around 95 percent.

Clearly, levels of lactose intolerance much more closely track cultural practices than they do anything we could call "race," and the fact that milk consumption by adults can only date back to the domestication of cattle, well under ten thousand years ago, suggests that the limited regional dominance of the genotype conferring lactase persistence is of recent origin. As a result, many believe that the alleles for this condition have been under intense positive selection in dairying humans—indeed, to some of the strongest selection that has ever been demonstrated in human populations; and although some doubts persist about the reliability of methods of measuring within-population selection, it is clear that cultural practices as well as other influences may help mold the genetic makeup of human populations entirely independently of their genealogical origins. To conclude the matter, the molecular geneticist Sarah Tishkoff and her colleagues recently showed that lactase persistence evolved separately in different human populations around the world, quite precisely tracking the cultural milk-drinking stimulus.

The foregoing examples are just a couple among many, but they do make it clear that both populations, and the individuals within them, are bounded by various definable biological limitations. On the individual level, no matter how highly motivated you are, you will never make it as a star basso profundo if you don't have the requisite lung power. And on the group level, this particular example immediately brings to mind the fact that some human populations display singular physiological capacities compared to others. Thus some groups of humans have acquired what appears, to us lowlanders, to be preternatural ways of coping with the problems of living at high altitudes and low oxygen levels. Perhaps the best-studied high-altitude people are the Quechua-speaking Indians of the altiplano in Peru. The altiplano is a high plateau that reaches elevations of fourteen thousand feet, well over the altitude at which Federal Aviation Administration regulations make it mandatory for pilots to have an oxygen supply, and it is a highly challenging environment for aerobic organisms of all kinds. As far back as the sixteenth century, the Spanish conquistadores were vigorously complaining of headaches, sickness, and muscular weaknesses up there on the altiplano, the nerve center of the vast Inka empire—even while, all around them, the natives were going about the rigorous business of their daily lives apparently unperturbed. How did (and do) the Quechua-speakers do it? Numerous studies have shown that high-altitude populations in the Andes cope with oxygen stress by having raised levels of hemoglobin in their blood compared to populations from lower latitudes. Hemoglobin is the molecule that makes blood red and that carries oxygen around in the bloodstream for release to the tissues. Additionally, along with their enhanced oxygen-transport capability, high-altitude Andeans exhibit "barrel chests," reflecting greater lung capacities, while also showing reduced tendencies to hyperventilate in compensation for low oxygen pressure.

Beyond this, population studies have also revealed that, while babies born to both Spanish and Quechua mothers showed reduced birth weights at altitude, the difference was greater for the Spanish babies. This suggested that not all of the high-altitude responses of the Quechua were entirely physiological, that is to say, simply a reaction (as barrel chests seem to be) to residence at high altitude since birth. Instead, some of the high-altitude population features, notably the increased oxygen-carrying capacity of the blood, seem to have a genetic component as well. If this deduction is correct, here we have a genuine population adaptation.

However, it appears that there are more ways than one to skin this particular adaptive cat. Residents of the high-altitude Tibetan Plateau, for example, appear to have found their own solution to dealing with the demands of a low-oxygen environment—a solution entirely different from the one adopted by the Andeans. Researchers found not long ago that, to allow for high energy expenditures in the prevailing low-oxygen environment, these people don't carry more oxygen in each red blood cell. Instead, they take more breaths every minute. To the same end, they also synthesize a relatively high amount of nitric oxide from the high volumes of air they take in. Nitric oxide is the gas that makes Viagra work, and it does so by dilating the blood vessels. Since wider arteries permit greater blood flow, increased oxygen reaches the body tissues in these people, even though the oxygen-binding capacity of each individual red blood cell is no better than normal.

The debate is just beginning over whether this remarkable mechanism was present in the original immigrants into the Tibetan Plateau or whether it evolved there over time. However, the fact that two different populations in different parts of the world have reached such dissimilar solutions to the same problem argues strongly that we are not looking entirely at physiological responses here, but at different heritable solutions. This conclusion is reinforced by even more recent work in Ethiopia, which has found that highly active residents at eleven thousand feet neither increase their hemoglobin counts as the Andeans do, nor breathe more rapidly like the Tibetans. They don't seem to synthesize more nitric oxide either: they have found some other way of nourishing the oxygen-hungry body tissues. As yet, nobody knows how.

Once again, then, we are faced with the omnipresent and entirely mundane fact that both individuals and populations vary, for reasons having to do with both the environment and with heredity. And while you might not be surprised that a group of Africans has reached a distinctive, if as yet uncharacterized, solution to the problems of living at altitudes where you or we would find taking each step an effort, it is all the more remarkable that Andeans and Tibetans, whom we know to be members of the same great branch of humanity, should have arrived at such different answers to the same predicament. Chances are that although the challenges they faced were significant (high plateaus are wonderful virgin territory for humans, but they're virgin for a reason!), the answers they found to the problems of living at high altitudes were based simply on random genetic innovations that arose early on in the populations concerned.

But humans are not simply physiological creatures; they are, supremely, cultural ones as well. Their possession of culture can help them accommodate to an incredibly wide range of circumstances. Tropical in origin, fully modern *Homo sapiens* definitively emerged from Africa only around fifty thousand years ago. Yet by twenty-two thousand years ago they were already at the Arctic Circle, in Siberia—at almost the peak of the last Ice Age! This was territory even the "cold-adapted" Neanderthals had shunned— by thousands of miles. We don't know what kind of physical adaptations those Ice Age humans might have acquired that allowed them to colonize such rigorous territory (for example, some human groups that live or lived in environments that are at least intermittently very cold have networks of tiny blood vessels that enable them to reduce heat-wasting blood circulation to the limb extremities). But whether or not the Upper Paleolithic Siberians had any such specializations at all, we can be absolutely sure that the main accommodations they made to their harsh habitat were cultural ones. Human beings are above all cultural creatures, and this—along with their very recent origin—is quite plausibly why they show so few physical variations that we can confidently chalk up to environmental response.

SO, WHY RACES?

Apart from trying to provide a historical perspective, the major effort we've made in this book has been to show that the local physical differentiation of populations in any widely distributed species is an entirely routine process in evolution and is indeed an essential one. Without it, macroevolution could never obtain a grip. At the heart of differentiation lie the twin processes of adaptation and random drift—which in itself come in various guises. And historically, people have sought adaptive explanations for differences among the major geographic groups of human beings. Classical writers attributed racial differences to the effects of climate, as did their successors right up until the time when Europeans began to colonize large swaths of the tropics, and once colonization was under way it began to be noticed that, beyond tanning a bit, the expatriates didn't change much in reaction to the new conditions. More likely, they (or their hosts) would simply fall victim to diseases to which they had no resistance. As a result of this lack of evident physical response to climatic stimuli, scientific attention ultimately turned to functional or evolutionary factors that might potentially have contributed to the physical differences among local populations. But even after the expenditure of much effort and ingenuity, the better part of a century later those inquiries haven't turned up much.

For example, it turns out that few if any of the classical physiognomic factors by which we recognize different regional human groups appear to have a huge amount of adaptive significance. Why do eastern Asians tend to have an "epicanthic fold" that often obscures the upper eyelid? Nobody knows for sure, but in all likelihood this trait was simply present in the small founding population, as a result of random variation. No surprise there: many traits simply hang around in populations for no better reason than they don't get in the way. Why do some populations boast broad, flat noses, while others have narrow, protruding ones? The nose is a major portal to the respiratory tract, and reasonably enough, the answer to this question has generally been sought in the different demands on the respiratory system made by different environments around the globe. In some parts of the world you need to humidify and warm the cold, dry air you breathe in before it hits the delicate lung tissues, while in others you don't. No studies have ever managed to show for sure, however, that the kinds of variation we perceive in nose shape make much practical difference in terms of the flow of air or in its preparation for the lungs. Indeed, it's not even clear that the huge noses of

Neanderthals had much, if anything, to do with the generally cold climates in which they lived. Quite likely, though highly diagnostic of their possessors, those striking Neanderthal schnozzolas were pretty much neutral in adaptive terms. And so on with almost any trait you might want to mention, the primary exception being skin color. Here we can make a pretty convincing case for adaptation, at least at certain points in the human past.

Human beings are by derivation tropical animals. Our African precursors lived in a range of generally low-latitude environments of high solar radiation; and solar radiation is hugely damaging to a wide variety of substances. Such substances include not only the materials from which hot-air balloons are made (balloons operating in high-altitude areas have to be replaced much more often than those flying close to sea level) but also human skin. The skin is our largest organ, and it is perhaps the most important organ of them all, playing innumerable roles in the definition, protection, and maintenance of the rest of the individual. But the skin is a delicate tissue that itself requires protection from the sun's rays. In naked-skinned humans, such protection is provided by melanin, a dense pigment that is manufactured in the outer layer of the skin. High concentrations of melanin in the epidermis may be genetically programmed or, within limits, may be stimulated by ultraviolet radiation (UVR), resulting in suntanning.

Just when our lineage of hominids lost the coat of body hair that their ancestors undoubtedly possessed is not known for sure. On physiological grounds it seems reasonable to suppose that the lean and hungry *Homo ergaster* had already shed it, but a recent molecular estimate (based on the MC1R gene you're about to read about) came in at only about 1.2 million years ago. Whatever the exact case, when hominids emerged from the shade of the African forests and shed their primitive hairy coat, their unprotected integument would have been highly vulnerable to solar radiation. Quite probably, that skin was already at least somewhat pigmented, and researchers have concluded, almost certainly correctly, that the gene variant responsible for dark skin pigmentation in humans was under strong selection at some time in the past, plausibly right back to the beginning of the tenure of *Homo ergaster*. In any event, it is clear that that there was every reason for the skin of all our African ancestors to have been heavily pigmented, and for high levels of pigmentation to have been maintained as long as those ancestors remained in their continent of origin. Sunburn alone is highly debilitating, and not for nothing do we find the world's highest levels of skin cancer in Queensland, where pale-skinned Australians are wont to wear minimal clothing in tropical conditions (and the indigenous people are sensibly dark).

Today, however, variation in human skin pigmentation around the world is amazingly extensive. There are as many people around with light skins as with dark ones—not to mention every possible intermediate. And again, we can plausibly look to environmental causes for an explanation. This is because for the early modern *Homo sapiens* who colonized the temperate regions of the globe, the adaptive calculation was entirely different from the one that had applied to their equatorial forebears. Here's why. It has long been known that the skin's relationship to UVR is a complex one, because this radiation both inhibits and stimulates the production in the skin of some chemical products that are very important to maintaining physiological balance in the body. Longer-wavelength ultraviolet (UVA) destroys folate, a vitamin in the B-complex that is essential to the proper production of DNA in the cells. On the other hand, the shorter-wavelength UVB promotes the synthesis in the skin of vitamin D, which is needed for the proper working of calcium metabolism. As delivered by the sun, these two kinds of UVR come in a package, and the anthropologist Nina Jablonski has very nicely characterized the resulting delicate minuet in her recent book *Skin*. As she points out, the necessity of balancing the inhibitive and stimulating effects of UVR upon the skin has led to two distinct gradients in environmental influence. One cline runs toward the equator from the poles, and the other goes in the opposite direction, but both act to favor darker skins in the tropics and lighter ones at higher latitudes. As a result, as early modern humans moved north out of Africa, they encountered strong selection against high melanin levels in their skin.

This of course makes a great story, of the kind to which our reductionist human minds seem innately attracted, and the need for heavy pigmentation in the tropics is self-evident. Still, perhaps there is simply a physiological or other cost to maintaining high melanin levels in the skin, and what we are seeing is simply the result of a relaxation of selection for dark skin pigmentation at higher latitudes. No matter: the theoretical expectations that emerge either way are generally borne out by measurements of skin pigmentation among native peoples at different latitudes. The correlation isn't perfect, but it is actually quite startling how well the expected pattern of lighter skins among native peoples at higher latitudes holds—especially given potentially confusing factors that include the short timescale involved in the human diaspora; the advanced cultural capabilities of the first modern human émigrés from Africa that helped them accommodate to new conditions by, for example, clothing themselves; and the ways in which populations have moved quite irregularly over the surface of the globe, at least with respect to latitude. The pattern applies best, of course, in the Old World; human beings have been in the Americas vastly less long. Intriguingly (if unsurprisingly) we now have molecular evidence that *Homo neanderthalensis* also fit this pattern of skin pigmentation, quite independently. In a recent DNA study two different laboratories detected a gene variant responsible for red hair and pale skin (the MC1R gene you'll read about in a moment) in Neanderthal DNA extracted from bone samples found at two different sites in Europe. However, the genetic variants that may have produced light skin color in Neanderthals are not the same variants that are at work in *Homo sapiens*—providing yet another example of how adaptive traits may not be reliable markers of population history, and pointing to the ease with which independent traits such as skin color can arise.

Interestingly, in looking at the molecular basis of skin color variations in humans (and in Neanderthals, for that matter), we can see the opportunistic nature of evolution at work. One of the genes influencing human skin color was actually first identified in zebra fish. These are creatures that normally bear the dark stripes their name suggests but that are "golden" (basically, transparent) if they possess a mutant form of a specific gene that influences pigmentation in the skin. And the researchers who identified this gene also showed that it was incredibly similar to its counterpart in humans. Indeed, when they injected the human form of the gene into golden zebra fish, the stripes came back! The gene concerned, cumbersomely known as SLC24A5, produces an enzyme that shows great variation among populations around the world, but its effects are not simple. A single amino-acid difference along the protein chain of this enzyme separates 98 percent of people of sub-Saharan African descent from 99 percent of Europeans. Almost all Africans have alanine in this key locus, while as many Europeans have threonine, and this accounts for a large part of the differences in skin color between the two populations, dark in the former and light in the latter. However, the same study revealed that some 93 percent of (dark) Africans and (much lighter-skinned) East Asians share the alanine allele; and this is good evidence not only that genes other than SLC24A5 must also be involved in determining skin color differences but that light skin color in Europeans and East Asians must be achieved by different genetic pathways. In other words, from a dark-skinned and presumably tropical common ancestor, each group must have independently arrived at a solution for living in higher-latitude circumstances.

The same researchers also looked at Caribbean peoples of mixed African and European ancestry and showed that skins were lightest among people homozygous for the threonine allele and darkest for those homozygous for alanine. In the end they concluded that up to 38 percent of variance in skin color was accounted for by differences in the SLC24A5 gene and that the balance was accounted for by other gene complexes. Several lessons are embedded in these findings, of which probably the most important is that, in the words of Stanford University's Gregory Barsh, skin-color variation should "not be confused with the concept of race." The darkest-skinned peoples are linked by environment of origin, not by genealogical descent. The findings also support the notion that skin color in *Homo sapiens* was initially dark, in response to strong protection from UVR, and that the lighter coloration of higher-latitude populations was independently achieved as a result of adaptation to an entirely different set of environmental imperatives. However, it has also been noted that there appears to be a "dominance component" for lighter pigmentation in certain genes, including SLC24A5, in which one ancestral guanine base was replaced by an adenine, resulting in reduced melanin production. Quite possibly, the tendency is for skins to lighten in the absence of strong selection to remain dark.

Hair color is another one of those interesting aspects of visible human variation that is correlated with latitude. Red hair has been shown to be controlled by a gene called MC1R, which affects a receptor on the surface of melanocyte cells (those cells controlling pigmentation). If you have two defective MC1R alleles, you will have red hair, you won't tan, and you'll be more susceptible to melanoma. However, there are about seventy-five different alleles for this gene in *Homo sapiens*, meaning that the probability that two randomly chosen redheads will have the same allelic combination is nowhere near 100 percent and is, in fact, much lower. The intensity of natural selection in the context of UV radiation has apparently promoted many ways to be a redhead (remember the Neanderthals).

Skin and hair color are not the only feature in which some correlation can be demonstrated between human physical attributes and latitude. There is also a tendency among human populations to conform to a pattern of body proportion that is quite generally seen among mammals: humans from tropical latitudes tend to have slenderer bodies and longer limbs than their counterparts from cooler regions. Once again, the explanation is pretty straightforward: tropical and polar mammals, human and otherwise, have completely opposite problems of body heat regulation. In the tropics you need to shed heat, which you can do by maximizing the surface area of the body available for radiating metabolic warmth. This is achieved by a linear build. In cold climes, on the other hand, the key is to preserve body heat, which you can do by maximizing body volume relative to surface area. For this, a rotund figure is ideal. And while once more the correlation isn't perfect, you do find the world's tallest and slenderest people in the equatorial regions, while arctic dwellers tend to be relatively short-limbed and to have plenty of subcutaneous body insulation. What's more, the contrast holds best for those peoples who have been where they are for the longest periods of time.

Still, once you stray beyond these fairly obvious features the search for adaptive differences begins to falter. This is particularly so when you look at potential cognitive variations within *Homo sapiens*. We all know that individuals of the same population show plenty of differences in cognitive abilities of all kinds—and there are many indeed of them. And it turns out to be tough to measure those differences in any meaningful way. It is harder yet to compare their significance—how do you rate mathematical ability relative to artistry, or street smarts, or a knack for fixing things? To put it another way, when an electrician arrives at your home to correct a hazardous condition it would be pretty useless to worry about what race he is; if you have any sense at all, your major concern will be about whether he's competent or not. And his "race" will be a pretty poor predictor of that.

To put the matter in perspective, none of this is very surprising when you consider that our brains are all-purpose machines, rather than instruments exquisitely burnished to do particular jobs. Evolutionary psychologists have suggested that the brain is a "modular" structure, with different parts (functional "parts," not necessarily anatomical ones) molded by natural selection to perform optimally in different areas of cognitive function. But in reality the human brain is a gloriously jury-rigged affair, formed by accretion over hundreds of millions of years, in which some very new structures communicate with each other via some very ancient ones indeed. No engineer would ever have designed a brain like ours—which is probably precisely the reason why it shows its remarkable emergent qualities. There is no quality control expert out there ensuring optimum performance in all respects, so there's little wonder that our brains show such a dazzling variety of combinations of aptitudes and obtusenesses. What's more, our brains seem to be almost infinitely malleable, molding and remolding themselves as a result of individual experience. And despite many claims made over the years, in complex societies such as ours, stratified in a thousand different ways—mostly economic and cultural, but including subtle influences like exposure to toxic substances—nobody has yet been able to figure out a satisfactory way of separating out potential heritable from environmental effects on *any* measure of cognitive ability.

Comparisons among populations from different places are even more difficult because they are invariably confounded by yet more profound cultural divergences than the ones you find even within hugely heterogeneous populations such as that of the United States. As we've just hinted, the cognitive capacity that underwrites the unique human spirit that we all share is a general-purpose ability, one that can be turned to many different purposes and ends. And as we saw earlier, it is an *emergent* quality, one that is supremely unlikely to have been molded by natural selection in the same way that we can reasonably suppose skin color has been. Much as we would like to think otherwise, we were clearly not fine-tuned by evolution, over vast eons, to be the extraordinary cognitive entity that we are. In which case, whatever the essence of "intelligence" may be, it makes most sense to view any variations we observe in it as "noise": simply the inevitable scatter around the mean that you'd expect to find in a species as biologically variable as ours.

CHAPTER 5

RACE IN ANCESTRY, FORENSICS, AND DISEASE

WE HAVE ALREADY covered a lot of ground discussing genes as they relate to human origins and diversity in various contexts. But we haven't finished yet, for race and genes have also figured widely in recent discussions of medicine, individual ancestry, and forensics, three areas that we've so far only touched in passing, yet are the focus of enormous popular and scientific interest. All three of these endeavors have been the target of large-scale government-funded or privately funded scientific projects, and parts of each are also embedded in mainstream biological research on the human genome. In addition, these endeavors have attracted commercial interest, making them not only highly visible in the popular press but also quite lucrative as approaches to a general understanding of human genetics. Each of the three areas we examine in this chapter is complex, and a simple "use it or lose it" stance on any of them would be short-sighted.

The use of race in understanding disease is a particularly pointed subject. Why wouldn't we want to use any tool possible in the attempt to characterize and stop disease? And if we are to use any approach possible, then the use of race in disease studies should certainly not be off-limits to researchers if it has utility and particularly if it is the only way to get the job done. Preventing and reducing crime is an important aspect of modern social life, right? Why not use every tool available to make society safer? And an individual's heritage and ancestry can be an important aspect of that person's psyche. Knowing where you come from, to whom you are related, and what culture you identify with are very important aspects of anyone's psychological makeup. Especially if you happen to be a member of a disenfranchised group or a member of a community that has experienced historical events involving forced dispersion or diaspora. Why shouldn't we use genetics and race to approach these important aspects of modern life? We have tried to make it clear throughout this book that our interest here is strictly in the scientific aspects of the concept of race, so we will not delve too deeply into the ethical or political issues involved in these three areas, as interesting and important as they are to the subject. And while some readers might feel it is impossible to separate ethics and science in this arena, we feel that a cogent examination of the scientific aspects of the use of "race" in modern biology can add greatly to clarifying our understandings of the ambiguities this concept raises.

So our purpose in this chapter is to examine the scientific advantages and disadvantages of using race in the three areas we have singled out. As with the rest of this book, we focus almost entirely on the science involved in these approaches and make our assessment of their validities based purely on the science. In doing this we address three critical aspects of the science concerned. First, by explaining what is involved in the approaches we examine, we hope to demonstrate the scientific strengths and weaknesses of using the notion of "race" in these endeavors. Second, we hope to examine how useful "race" has been to the overall goal of each of them. And finally, we ask whether approaches employing the concept of "race" are necessary to the accomplishment of the goals of the three endeavors or whether the use of race may simply be a convenience that can be replaced by some other approach.

In doing this we are not advocating the immediate trashing of the use of stratification of population along racial lines in any of these three areas, although we do feel that this would in fact be for the best. We are simply pointing out that while race-based approaches may be common today, we most likely will be using other approaches that do not involve the concept of race tomorrow. There are three reasons to draw this conclusion. First, we feel that the use of race in these three areas is sometimes flawed theoretically; and if the use of race is deficient on first principles, then we probably should stop using it. Second, we hope to show that there are alternative ways that do not involve race-based approaches to accomplish the goals of each of the three endeavors. Even if these alternative approaches to using race are more complicated, costly, and time-consuming, we should probably switch gears and start developing and using them. Finally, and this has been an overriding thesis in this book, because of the continuing reintegration of human populations with each other, race-based approaches simply will cease to be useful before long. Rather than pin our hopes on a concept or approach that will be of limited use, for a few more decades at most, we should be thinking about the long term and about approaches that do not rely on race. Our hope is that scientists will speed up the search for nonracial approaches to the understanding of disease, ancestry, and forensics as much as they can. And we begin our discussion with the most complex and difficult area to understand—race and genetic disease studies.

RACE, GENES, AND DISEASE

The discovery of disease genes and the epidemiological aspects of disease is an important endeavor in modern science. Much of the public appeal of the human genome project has derived from the promise of medical advances as a result of the genomic knowledge obtained. The rationale for this is that if we know which genes are involved in a disease we will understand its etiological basis and can better devise methods for early detection and treatment. A major argument made by well-meaning biologists and medical scientists in favor of retaining race as an organizing principle in such research is that race can help a lot in understanding the occurrence of disease and is thus potentially an important tool in the discovery of the genes causing particular diseases. To these scientists, race is therefore also potentially important in the development and application of cures and treatments for human ailments. Since we see this argument over and over again in the medical literature, we need to look more closely at how race has been used in attempts to understand disease and at the potential of this approach for devising cures to genetic disorders.

In order to do this, we first need to understand the genetics and research approaches in human genetic diseases. This understanding requires some discussion of the genome itself and a look at how it is analyzed using single nucleotide polymorphism (SNP) approaches and how classical principles of genetics such as linkage and linkage disequilibrium are used in human genetic research.

HUMAN GENOMES EVERYWHERE

Let's start with the fact that, while both mtDNA and Y chromosomal DNA harbor genetic changes that are implicated in disease, the great majority of DNA-related diseases that impact humans are found scattered throughout the rest of the genome, that diploid part of the genome, that we have yet to examine in this book. Of course, it goes without saying that we should do as much as possible to increase the health of people on this planet. In fact, as we've already emphasized, if *any* tool exists that has the potential to advance medicine, its use demands to be considered seriously. And this, of course, raises a set of important questions. First, we should ask whether the approaches and techniques that use race as a foundational assumption are necessary or whether they simply present themselves as convenient supposition in the prevention or treatment of disease. If the notion of race is in fact a convenient rather than a necessary one, we need also to ask whether there are other, more accurate, approaches that do not need race as a founding assumption. And the larger question, of course, is whether "race" has in fact made a difference in medicine or whether it ever can. In asking how race is used in examining human disease, we first need to realize that part of the answer to this question depends on how we understand human genomes in the first place.

The haploid human genome has over three billion base pairs in it. Recent sequencing of dozens of individual human genomes has revealed considerable differences at the level of base pair changes and at the level of overall content. While this book was being written, over a six-month period, we had to revise this section three times to account for rapid advances and new reports of the sequencing of whole human genomes, the so called "personalized" genomes. Because the sequencing technology involved has entered into what the technophiles call "next generation sequencing," whole genomes and proxies for whole genomes are being generated at an unprecedented pace, for relatively small amounts of money. Our first draft of this book reported that seven genomes had been sequenced (four individuals from Asia, two Caucasians, and two Yoruban individuals from Africa). The two Caucasian genomes and one of the genomes from Asia are those of three famous scientists, Seong-Jin Kim, James D. Watson, and C. Craig Venter. The other four came from unidentified donors. Our second draft added that five genomes from African people (Khoisan and Bantu) had been produced. These expanded genomes included the San individuals named !Gubi, G/aq'o, D#kgao and !Ai (the click sounds involved in pronouncing their names are difficult to represent in conventional text, hence the symbols !, /, and #) from Namibia, while the genome of Archbishop Desmond Tutu represented the Bantu heritage.

In our third draft, we had to include five more genomes that were derived from two family studies of inherited diseases, and we'll discuss these in some detail in a moment. In one study, researchers at the Institute for Systems Biology in Seattle sequenced the genomes of two children affected by Miller syndrome (causing craniofacial alterations) and primary ciliary dyskinesia (PCD, a disorder that affects organs in the body dependent on the action of cilia), plus those of their parents. In a study accomplished by the Baylor University Human Genome Sequencing Center, the genome of a scientist named James Lupski was sequenced. Dr. Lupski has Charcot-Marie-Tooth syndrome (which causes severe problems with the peripheral nervous system). In addition, the "1000 Genomes Project," which seems to have the same moving-target problem that we had in writing this section of the book, promises to have another two thousand (so why not call it the "2000 Genomes Project"?) human genomes sequenced at what is called "low coverage" by the end of 2010. Low coverage simply means that the sequencing intensity is not as high as in the studies we mentioned earlier.

So, what about the variation that is observed when these genomes are compared? There are many ways to generate a number that provides a measure of variation in these genomes. The simplest is to compare the sequences from each of these individuals to a reference genome, and, fortunately, such a reference genome exists under the name of the "human reference assembly." This reference sequence was the product of the original Human Genome Project. The number of sites in the fully sequenced genomes we list above that are different from this reference assembly ranges from 2.5 to 4 million bases, reinforcing the initial suggestion that humans are on average 99.9 percent similar to each other at the level of the genome. We can also ask how many differences there are between any two or three individuals with "personalized genomes." For instance, when we compare the genomes of one of the anonymous Chinese individuals (YH), J. Craig Venter, and !Gubi, we see that, of the on-average 3 million variable positions in each of the three genomes, 741,000 positions are the same in all three individuals, 452,000 are the same in Dr. Venter and !Gubi, 509,000 are shared between YH and Dr. Venter, and 530,000 are shared between !Gubi and YH. Of the remaining variable positions, 1,036,000 are unique to Dr. Venter, 958,000 are unique to YH, and a whopping 2,038,000 are unique to !Gubi. The large number of variable positions unique to !Gubi turns out to be a general characteristic of the Yoruban, Bantu, and Khoisan genomes so far analyzed. When all of the personalized genomes are compared together (which reduces the number of unique differences amongst the individuals), !Gubi has over 700,000 unique variable positions, Dr. Venter over 160,000, and YH over 80,000. The larger number of unique variable positions in !Gubi's genome clearly reflects the historical depth of his genome. And, as we will see later, the higher degree of variation in genomes like !Gubi's requires special measures when examining disease at the genomic level.

SNPs

We need to keep in mind that only a small proportion of these variable base pairs codes for gene products. Slight differences in sequence and arrangement of DNA are, nonetheless, scattered throughout the genomes of humans. If they involve a single nucleotide change, these variants are called "single nucleotide polymorphisms," or SNPs (pronounced "snips"). Each SNP position can have either a G, A, C, or T in it, and each variant at a SNP position is called an allele for that SNP. Collections of SNPs that are in neighboring regions of a chromosome are grouped together into what are called haplotypes. So every SNP position in your genome can have at most two alleles in it and more than likely only one. In the former case, you are called a heterozygote for that SNP, and in the latter, you are called a homozygote for that SNP. Other variants can involve differences in the lengths of short stretches of DNA caused by variation of short two, three, or four base repeats called microsatellites and also more drastic changes like chunks of chromosomes being rearranged.

Think of a chromosome, or a region of a chromosome, as a string of beads (this is somewhat of an oversimplification because a lot of the variation in the human genome is caused by rearrangements of our chromosomes). Each of these beads represents a SNP and each is a different color, depending on which allele your SNPs have. Organisms like us that reproduce sexually get two strings of beads, one from their mother and one from their father. If we are looking at identical twins, then the strings of beads will be identical in the order and the color of the individual beads (identical in the order of SNPs and in their alleles). During the production of eggs by a female, in meiosis when the cell divides, the beads on one strand will randomly exchange with the beads on the other strand. As a result, the string of beads passed on to the offspring by this female will not look exactly like the original strings bequeathed her by her parents, though they will still have all of the same kinds of beads that her mother and father gave her. The same thing happens in males with respect to their sperm. If there were no variation—if there were no SNPs—then all of the strings of beads would look the same, and all people would be genetically identical. In other words, if there were no exchange of beads between strings, the strings would look different from one person to the next but they would be the same from generation to generation within the population. And this, of course, would make disease correlation studies really difficult. It is the interchange of beads from one string to the next that makes studies where a disease is associated with a SNP or group of SNPs possible in practice. So how, exactly, does this work?

To date, over 15 million SNP positions have been found amongst the sequences in human genomes. This doesn't mean that if we could compare their genomes Angelina Jolie and Brad Pitt would turn out to show 15 million differences between their DNA sequences. Fifteen million is simply the number of SNP variants found in *all* of the hundreds of comparisons of humans done so far. Instead, we would almost certainly find that Brad and Angelina would have around 3 million SNPs that differ between them (ignoring the Y chromosome, of course, which Angelina certainly doesn't have).

SNPs have become the major markers in disease studies, just as mtDNA and Y chromosomal DNA have become the classic markers of human movement across the globe. And they have been mapped in large numbers of humans via massive sequencing and resequencing projects. The "HapMap project" was designed specifically to assist medical scientists in associating genetic loci with specific human diseases using these SNPs. The HapMap consortium is quick to point out that the immediate goal of their project is not to identify the disease-related genes themselves but rather to provide a central repository of SNPs and other genetic markers to make them available for what have been called "association studies." The HapMap project stores these SNPs as *in silico* markers in a computer-based repository at http://hapmap.ncbi.nlm.nih.gov/.

Remember that more than fifteen million SNPs have so far been discovered as of the writing of this book, using high throughput sequencing and resequencing methods. This is a huge number to deal with when doing a disease association study, so commercial outfits have taken the results of the HapMap project and have reduced the fifteen million SNPs to approximately one million SNPs that are in practice most useful for disease studies. They can do this because many of the SNPs are actually linked to each other very tightly and therefore represent "semi-redundant" information.

NEEDLES IN HAYSTACKS

Trying to find a gene that causes (or even partly causes) a disease is like trying to find the proverbial needle in a haystack. If we define a gene as a DNA sequence that is translated into a protein, then in fact some causative factors may not even reside in genes but rather in areas outside of protein-coding regions, in stretches of the DNA that might instead control the *expression* of genes. This of course complicates matters yet further when we are looking for "disease genes," because it means that there are in fact two stacks of hay—one for areas of the genome that code for proteins and one for areas of the genome that are outside of those genes. In the days before geneticists could rapidly, inexpensively, and accurately type large numbers of SNPs for research on disease genetics, they used an approach called "candidate gene association studies." This focused mostly on the stack of DNA sequences coding for proteins. If a researcher was, for instance, studying a disease that caused a malfunction in the red blood cells of individuals, and could further narrow down the cause to the proteins that make up hemoglobin in red blood cells, then the genes coding for the proteins that make up hemoglobin would become "candidate genes" for association with the disease. This is what geneticists call a "hypothesis-driven" approach. And it requires that the geneticist know a lot about the disease and the proteins potentially causing it. As Javier Simon-Sanchez and Andrew Singleton put it, "genuine success required sufficient understanding of the disease process to enable selection of the correct gene, the right variants to type within the gene, and—most importantly—the presence of variability within the gene that could alter function or expression." Because of the huge number of genes (approximately twenty-five thousand in humans) and the even larger number of variants for those genes, the probability of validly testing hypotheses about candidate genes was extremely low. Consequently, this approach has had only very limited success, and for the most part it is not considered very valuable for understanding genetic disorders in humans. Still, it has been found successful for certain disorders such as hypertension and metabolic disorders. Without the candidate gene approach we might be lost, but the good news is that there is a huge amount of genetic variation in human populations that could possibly be associated with diseases and that geneticists have developed a variety of tricks for using this variation in disease studies.

THE DISEASE/VARIANT SPACE

To understand how this genetic variability is used by scientists to detect genes for disease at the whole genome level, and how race comes in to all this, we need to think about how the variation is distributed in populations and, more importantly, how the alleles that are relevant to diseases are distributed. To explain the importance of human variation for correlating disease to the genes in your genomes, we need to introduce an idea that correlates the occurrence of variation with the frequency of occurrence of disease. One early hypothesis about how all of this gets tied together was proposed by David Reich and Eric Lander and is called the Common Disease/Common Variant (CD/CV) hypothesis. This idea predicts that risk at the genetic level for a disease that is classified as common will be due to disease alleles that exist in the population at relatively high frequencies (usually around 5 percent). The implication of the CD/CV hypothesis is that any major disease-controlling locus should have one, or a few, very common disease-causing alleles at that locus. Reich and Lander have called this the "simple allelic spectrum" for disease loci. Most of the straightforward approaches developed for analyzing variation at the population level depend on the assumption that the allelic spectrum is simple in this way. These approaches include linkage studies using co-transmission of genes, association studies using linkage disequilibrium (which we will discuss in detail below), and the candidate gene approaches we just looked at.

In order to accommodate more complex disease patterns, in 2008 Mark McCarthy, David Goldstein, and colleagues introduced an idea that we call the "disease/variant space." In this way of looking at disease and variation, we plot the "frequency of a gene mutation" on the X axis of a graph and the "effect of a gene" on the Y axis. The space in the graph can roughly be divided into four quadrants, as shown in figure 7.

Quadrant I is filled with alleles that are found at "rare" frequencies (less than one in one thousand haplotypes) but that have large effects in the expression of a disease. Easily interpreted diseases inherited in Mendelian fashion lie in this quadrant. Conditions such as Huntington's Disease and cystic fibrosis are the prime examples of syndromes residing in this quadrant of the disease/variant space. Disorders like these are easily mapped using pedigrees of affected individuals and linkage analysis, an approach we will explain below. Quadrant II is where the variants for a disease are very common (greater than one in ten haplotypes) and where the effect of the gene on disease is also large (as in quadrant I). Such variants are extremely hard to find and to characterize, and for the most part they are considered almost nonexistent by human geneticists. Perhaps the most famous disease of this kind is sickle-cell anemia in Africa, where the frequency of the allele that causes the disorder is found in very high frequency in areas where malaria is prevalent, and the condition behaves as a classical Mendelian disease. The causative allele for sickle-cell anemia (a highly debilitating condition in the homozygotes that causes severe anemia in those affected as a result of the sickling of red blood cells) is maintained at a high level in the population because in a heterozygous state (one good copy, one bad copy), it provides a level of immunity against malaria. Quadrant III is the disease/variant space where the frequency of the allele is very

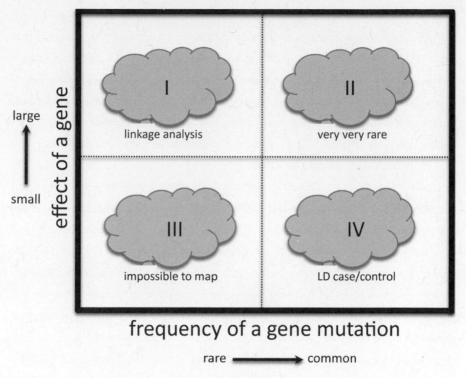

FIGURE 7. *Disease/variant space diagram (Goldstein et al., 2003) as discussed in the text.*

low and where the effect it has on a disease is also low. Such alleles, and how they are associated with disease, are essentially impossible to map using current tools because of the diffuseness of their effects on genetic disorders. It is, then, in quadrant IV that all the action resides with respect to current gene association technologies. This is where genes are found that exist in populations at high frequencies (from one in ten to one in one hundred haplotypes) and that have small effects on the expression of the disease. This region of the disease/variant space is where studies focus on large population-level studies of genes, and it is this part of the graph that is the target of the HapMap project and where race is most often used.

When the gene-candidate approach doesn't work the needle-in-the-haystack problem starts to worsen, because the entire genome then has to become the focus of the association study. Hence the term "genome-wide association study" or GWAS. The GWAS approach takes advantage of the opportunities offered by high-throughput approaches to SNP genotyping and of next-generation DNA sequencing. GWAS studies can be classified according to the approach taken to whittle down the haystack problem or according to what is called study design. In the first way of dividing GWAS studies, the approach can either involve using linkage or an unrelated approach called linkage disequilibrium (LD). While these two approaches sound alike, they are really quite different. GWAS linkage analyses ordinarily use a very simple analytical approach to pin down the co-transmission of a disease with a gene or chromosome region. Such analyses are called linkage studies because they rely on the actual linkage of a disease to a characterized genetic variant like a SNP.

How is linkage mapping accomplished? Let's return to our beads on strings and envision the following. In a family or small group of people, a particular disease occurs. When we analyze the genomes of the individuals who have the disease concerned, we discover that there are four different kinds of strings of beads (chromosomes), one each for the four individuals with the disease. Remember that the sequences of colors along the strings represent different haplotypes. There are also four strings of beads representing four offspring without the disease. The order of beads in the four offspring with the disease is CGCCCC, CGCGCG, GGCGCC, and CGGGGG, where C is cyan and G is green. The order of the beads in the offspring without the disease is CCCGCG, CCCCCC, GCGCGC and GCGGGG. Let's line all these up, and look at the pattern of the beads, where on the left are strings with the disease, and on the right are the strings without the disease:

TABLE 4. Example of linkage mapping discussed in the text (not to be confused with linkage disequilibrium)

Sick offspring	Well offspring
C-**G**-C-C-C-C	C-**C**-C-G-C-G
C-**G**-C-G-C-G	C-**C**-C-C-C-C
G-**G**-C-G-C-C	G-**C**-G-C-G-C
C-**G**-G-G-G-G	G-**C**-G-G-G-G

Note that the only position in the string of beads where there is a distinct difference between the two groups is in the second bead (underlined and bold). All of the other positions in the strings have both cyan and green beads in those positions. The second bead on each string is green in the sick offspring and cyan in the well offspring. Because there is a straightforward co-transmission of a green bead in this position in offspring with the disease, and a cyan one in those without, the region of the chromosome where this color-change occurs is where the gene or genes for the disease reside(s). Now, simply think of each of the beads on the string as a SNP on a chromosome, and the colors as being either a cytosine (C) or guanine (G) in the SNP position, and you have the situation that obtains in simple pedigree-based linkage studies.

This is an idealized case, and it has not been found to apply in the vast majority of the diseases for which genetic analyses have been performed, in which there is usually a less-than-complete differentiation between the sick and the well individuals.

GWAS, LD, AND RACE

If the allele disease spectrum does not allow linkage analysis, then the region in the disease occurrence space that gets the attention is quadrant IV—which is where Hap-Map has seen its greatest purported success. This is also the region of the disease occurrence space that requires the development of high-powered statistical tools to detect LD. Whereas linkage is quite easily detected using the rationale we mentioned above, LD is much harder to pin down. Its major differences from classical linkage analysis are that it doesn't use pedigrees and that it has a highly developed statistical context. Let's look at a population where a disease locus resides on a single chromosome and the population starts out with four ancestral chromosomes for this disease, as shown in the figure below. The symbols on these four chromosomes simply represent different SNP marker states. For this example a disease locus (D) that does not have a marker for it (after all, we want to be able to find it) and a special marker state on the chromosomes of diseased individuals called M exist. None of the other chromosomes have the special marker state M; instead, they have alternative states for that area of the chromosome.

After many generations of breeding and crossing-over amongst the chromosomes, the distribution of the markers in the population might look like the second set of eight chromosomes drawn in the bottom of the figure. Linkage disequilibrium essentially measures the statistical association of the marker state (M) with the disease (D). Why are they statistically associated? Because they are so close together on the chromosome that their disassociation by crossing-over will be a rare event. So LD in essence measures the probability of recombination changing the haplotypes in which both M and D are found. In some cases the marker SNP (M) actually resides within

FIGURE 8. *Schematic for the basics of linkage disequilibrium as described in the text.*

the disease gene (D). In this case, there is a complete correlation of the marker with disease.

Now let's look at a real example of LD that involves the determination of LD in a small part of chromosome 5 associated with the disease diastrophic dysplasia (a disorder involving problems with bone and cartilage development and causing short stature and shorter-than-average arms). Montgomery Slatkin described an LD study in Finland where a region of chromosome five was analyzed for two genetic markers. Instead of the complicated chromosomes we drew above, he described simple chromosomes that had haplotypes called 1-1, 1-2, 2-1, and 2-2, corresponding to the different states of the two markers. In a group of people without the disorder, the number of haplotypes found for 1-1, 1-2, 2-1, and 2-2 were 4, 28, 7, and 84 respectively. These individuals without the disease served as a control population for assessing whether or not the population of individuals with the disease showed LD. When individuals with the disease were examined, they had the following numbers for the four haplo-types—144, 1, 0, and 2 respectively. These numbers indicated a highly statistically significant association of the disease with the 1-1 haplotype. In other words, there was strong LD of the disease with the 1-1 haplotype. By measuring the degree of re-combination between the markers, scientists were also able to pinpoint the location of the two disease markers to within seventy thousand base pairs. Compare this to the linkage approach, which can only get the mapping down to within one million bases or so. The reason for this increased precision of the LD approach is that the LD stud-ies take advantage of many generations of recombination that have whittled away at the chromosome region where the markers and disease reside. This causes an increase in recombination breakpoints that get ever closer to the marker and the disease gene. GWAS studies based on LD therefore have the advantage of being more precise (about one to two orders of magnitude more precise); they don't require pedigrees; they can be accomplished on broad data sets that can serve multiple purposes (be used to ex-amine more than just one targeted disease); and they can be applied to disorders that don't necessarily occur in simple Mendelian fashion. In fact, complex disorders such as diabetes, asthma, and heart disease, and neurological disorders like schizophrenia and depression, have all been targeted using the GWAS LD approach.

So far so good, but where does race enter into GWAS studies that use linkage disequilibrium? First, in setting up the HapMap samples, scientists have used what they call "ethnicity" to designate which human genomes they have "scanned" to obtain the haplotype database. The original HapMap samples came from 270 individuals from the following self-identified "racial" groups—Yoruba (representing Africans), Japanese from Tokyo and Chinese from Beijing (representing Asians), and Caucasians in the United States (representing Europeans). Each time a sample was collected, a mother, a father, and a child were obtained in a "trio," so that there were 30 trios or 90 individu-als from each of the three major geographical groups. A later phase of the HapMap project expanded the number of people in the database to 1,300 and classified them in more specifically defined groups (see table 5).

Why these particular people? The different groupings of samples were chosen because they supposedly can be considered more or less closed breeding groups. But the overall reason these groups were selected is because of the assumption that closed breeding would ensure that more unique SNPs would be found close to the disease locus. As we've already seen, the mtDNA and Y chromosomal DNA agree that all

TABLE 5. Stratification groups in HapMap

Label	Population Sample
ASW	African ancestry in SW United States
CEU	Utah residents with northern and western European ancestry
CHB	Han Chinese in Beijing, China
CHD	Chinese in metro Denver, Colorado
GIH	Gujarati Indians in Houston, Texas
JPT	Japanese in Tokyo, Japan
LWK	Luhya in Webuye, Kenya
MEX	Mexican ancestry in Los Angeles, California
MKK	Maasai in Kinyawa, Kenya
TSI	Tuscans in Italy
YRI	Yoruba in Ibadan, Nigeria

humans on the planet today are where they ultimately are, following origin in and migration from Africa. Since the migrations have been both rapid and recent, it is no surprise to find that the majority of variation we see on the planet is, simply put, found in Africa and that the majority of variation that we see overall in human populations, no matter where they come from today, actually also exists in Africa. As the ancestors of contemporary non-African *Homo sapiens* started migrating out of Africa some sixty thousand years ago, they carried with them a great deal of this variation but not all of it. This means that there is some variation in African populations that is not seen outside the continent, simply because some of it did not move out of Africa with these migrations. The results of the whole genome sequences of !Gubi, G/aq'o, D#kgao, !Ai, and Bishop Tutu bear out this general reality of human genomes.

What happens in many cases is that the variation observed *outside* Africa is a subset of the variation *within* Africa. But another thing also occurs to structure this variation. Because the migrant populations were much smaller than the ancestral African populations that gave rise to them, they were more susceptible to genetic drift. This caused many novel haplotypes to increase or decrease randomly, and it has produced situations where haplotype frequencies have become quite different from one geographic region to the next. And yet another important thing has happened, as a result of the time and population sizes involved. We have noted that the linkage of haplotypes to each other as a result of their close proximity on the chromosomes is an essential tool for the HapMap project. But what has happened in African populations is that the larger populations, together with the longer time involved, have allowed haplotypes to get mixed up more than in other populations. So the linkage disequilibrium in Africa is much smaller than in other parts of the world. As a result, the choice of ethnic groups by the HapMap project, as a way to focus on data collection, is actually no more than a choice of convenience.

So what else does race add to the LD approach? Well, it turns out that if one is using a large population to search for linkage disequilibrium, the substructuring of the larger population might actually artificially result in the detection of LD. This will happen if one subpopulation has a really different allele frequency makeup from another. Here is how it works in an extreme case. Let's say that we have a large population of dogs, and we want to determine if there is LD for a particular pair of loci. But this large population of dogs is made up of both rough-haired collies and smooth-haired collies. It turns out that the rough collies are fixed (have the same allele at a particular gene) for two loci (the ESS locus and the ARR locus), and the alleles they are fixed for are called S and R. When we analyze the smooth collies for these same genes, we find that the smooth collies are also fixed—but for other alleles, that we will call s and r. The genotype of the rough collies is S/S, R/R, and they have only one kind of haplotype: SR. The genotype for the smooth collies is s/s, r/r, and they too have only one haplotype: sr. The two other possible haplotypes for this system, Sr and sR, just don't exist in either of the two types of collies. If we then combine the two kinds of collies into a single population (since after all, they are both collies), we will see that because of the way LD is calculated the $LD = 1.0$; in other words, there is perfect and high LD between the two loci.

But there is obviously a problem here. Because we have independent evidence that there are two subpopulations where there is absolutely no LD (if there is only one haplotype in a subpopulation, there can be no LD). For the smooth collies, the haplotype frequencies are $r = 1$ and $s = 1$, and for the rough collies the haplotype frequencies are $R = 1$ and $S = 1$. If we try to do LD analysis in either of these subpopulations, we find no variation—and hence linkage is not detected. And if the two subpopulations on their own don't show LD, then putting them together shouldn't create LD. This is, of course, an extreme example, but it shows that when two subpopulations have skewed allele frequencies at loci being tested for LD over both, the tests results should be closely scrutinized for the false detection of LD. The skewed allele frequencies almost always result from some substructuring of the population being studied.

Detection of LD is therefore impacted strongly by the presence of subpopulations. This is why during a GWAS, scientists need to examine their samples for substructure and eliminate or control for it. Hence, the HapMap Consortium divided their original overall sample of about 300 individuals into three smaller samples corresponding to Europe, Asia (China plus Japan), and Africa. While this particular stratification approach avoided the problems that substructuring poses when the entire data set is looked at, it did not eliminate the possibility that each of the four initial subpopulations was itself substructured. The only careful, logical, way to approach this problem is to stratify the individuals at the smallest subpopulation level possible or to avoid genes in which subpopulations significantly differ in allele frequencies. And in a third version the HapMap project expanded its samples to over 1,000 individuals from 11 populations. But are there even now enough populations to avoid LD?

Complicating matters is the existence of another set of human samples called the HGDP-CEPH (Human Genome Diversity Project—Centre d'Etude du Polymorphisme Humaine). This effort accessed its samples, and ascertained their origin, very differently from its HapMap counterpart. HGDP-CEPH is a human tissue culture collection of 1,063 cultured lymphoblastoid cell lines (LCLs) from 1,050 individuals belonging to 51 world populations. Panels of DNA are prepared from the cell lines and distributed as a means for doing population genetic studies on these groups. But as noted the samples were ascertained in a completely different way than the HapMap equivalents, and by comparing the population dynamics of the HapMap samples and the HDGP-CEPH samples, Yungang He and colleagues showed that the original HapMap stratifications were somewhat flawed. Specifically, the HapMap Chinese and Japanese populations, when combined, show the inflated LD problems we've already discussed. The upshot is that two of the four original stratifications were actually inappropriate for LD studies. And the two that were appropriate do not offer a broad enough representation of human beings to say anything about "race." Stratifying into more and more subpopulations might be one answer, but because of the large degree of admixture that is evident in almost all the populations we looked at earlier, the biological meaning of such subpopulations becomes less and less evident.

AIMs

An approach called admixture mapping actually takes advantage of the fact that humans like to interbreed with each other. This technique relies on the existence of "ancestry-informative markers" (AIMs), genetic markers that are found in high frequency in non-admixed ancestral populations. So instead of solving the problem by going to smaller and more cohesive entities as the unit of analysis, this approach seeks a solution by going to larger populations that have these so-called AIMs. But this leaves us with one of the more severe "chicken and egg" problems in human genetics. Where do we draw the line for optimal substructuring of a particular system? At three groups? At eleven groups? At fifty-one groups?

A colorful approach has been proposed for attacking this problem. Literally a colorful approach, because it uses colors to represent possible populations in a study and then graphs the probability that each individual belongs to a certain population. Its name is STRUCTURE. If there are indeed three substructured populations in a study, the desired result of a STRUCTURE analysis would be a graph with three purely, but differently, colored blocks that represent the three "real" populations. Using a pre-described model the algorithm allows the researcher to compute probabilities for any number of populations (K is used as the variable for number of potential populations) in a dataset. At the end of a STRUCTURE analysis, you have statistics for $K = 2$ populations, $K = 3$ populations, $K = 4$, $K = 5$, and so on. Under the conditions of the model and using various statistics the researcher can then determine which of the Ks is the best and infer how many "real" populations there are in the sample. But while this line of attack is promising for many kinds of animal populations and tells us that there are many more structured populations out there than we originally thought, what exactly is going on is not entirely understood. The anthropologist Deborah Bolnick has recently offered a particularly unsettling critique of this approach. Specifically, Bolnick points to three reasons why STRUCTURE doesn't work for human populations. First, the approach only offers an approximation of population structure. This problem exists because one of the inputs into the model can only be approximated, and as the authors of the algorithm suggest, "the assumptions underlying [this approximation] are dubious at best." Second, if the data set is complex, and human population data are extremely complex due to interbreeding, then separate runs of the algorithm with the same data set will give different results. Finally, the basic model that the algorithm is based upon may not be the most appropriate for the broad array of human population systems that exist. Being restricted to one model for deciphering structure of human populations is like trying to put a square peg into a round hole, and the STRUCTURE approach is like advising, "don't force it, just get a bigger hammer." Even if STRUCTURE is used, Bolnick suggests that the results "are consistent with our current understanding of human genetic structure," which is that human variation and human evolutionary history are peppered with admixture. Bolnick warns that when used on human populations this approach plays a negative "role in the reification of race as a biological phenomenon."

In any event, because of the severe "chicken and egg" problem involved here, and also because we suggest that we can do much better than the current HapMap approach and sampling regime, we are led to the conclusion that the notion of "race" really isn't helping much here. It certainly isn't necessary, and actually it can be quite inconvenient to use, not only for genetic reasons, but for social ones as well.

CONVENIENT OR NECESSARY?

In writing this book, one of the most common arguments we have heard from human geneticists who want to continue to use race is that "it is the best we have." But is this actually the case? It certainly isn't the only procedure available, as there are other study designs that have been suggested as useful for GWAS. Admittedly, the most commonly used study design is the "case-control" procedure that uses LD and race to stratify populations. As described above, many individuals with a disease (case) are compared to many individuals without the disease (controls). But we have just looked at how this approach works, and we have seen that while this design is the easiest of the options to carry out, it also makes the largest number of assumptions that need close attention when analyzing the data. If these assumptions are not recognized and dealt with, the study will suffer from biases leading to misleading inferences. The case-control approach requires that substructuring is either minimal or compensated for in the statistical design of the study. The HapMap has been used extensively in these kinds of studies, and, as we've pointed out, race is used in this approach to do two things. First, using race supposedly minimizes the false detection of LD by stratifying potentially substructured populations into the design of the study. Second, race increases the number of SNPs and perhaps individuals with SNPs that might be associated with a specific disease. Both of these aspects of race are imposed in order to increase the statistical power of a case-control GWAS study, but another major advantage of this approach is its rapidity. Specifically, large numbers of case and control individuals can be rapidly assembled, based on simple designation of study participants as having, or not having, a disorder. The assemblage of case and control groups is then examined for LD.

Another kind of study design for GWAS is called the trio design. This approach uses information both on individuals that are affected with a disorder and those individuals' parents. Only trios in which the offspring is affected, and the parents are not, are included, and all three individuals are genotyped. Because the genotyping will tell the researcher which SNPs are heterozygous, the frequency at which each SNP contributes to the affected offspring can be calculated. If there is no association of a disorder with a SNP allele, then the expected transmission of a SNP in a population of triads will be 50 percent (based on the luck of the draw during reproduction). On the other hand, if the SNP *is* associated with the disorder, it should be transmitted in excess of 50 percent in the affected individuals. Unlike the case-control studies, this design is not dependent on stratification because all that matters is parent-to-offspring transmission of SNPs and whether or not an individual has a genetic disorder.

A third approach to GWAS is the cohort design, which involves collecting loads of baseline information on a large number of individuals. These individuals are then observed to assess the incidence of the disorder in the large data set. The trick with this design is to place individuals into subgroups based on genetic variants and then to observe the groups (cohorts) for the incidence of disease. The problem with cohort studies is that they are very expensive and time-consuming. Some scientists suggest an additional disadvantage of cohort studies is that the disease must also show some variation in its impact and expression. However, it is actually possible to see this requirement as an advantage, in that it does not oversimplify diseases such as schizophrenia and bipolar disorders.

To be clear, the major advantages of using race in GWAS are that the stratification of data along racial lines makes the techniques rapid, the collection of baseline information is simple, and it yields enhanced statistical significance. All three of these advantages are of course desirable in studying disease. But while taking longer and requiring more baseline data, both cohort and trio studies make fewer assumptions, take a more sophisticated view of genetic disease state, and utilize the principles of genetic inheritance more precisely than case-control studies do. In addition, although case-control GWAS requires that the analysis take advantage of stratification effectively, we don't believe that this is necessarily the case in practice. Another problem with GWAS case-control design is that for technical reasons it is now easier to conduct such studies on people of European descent, leaving studies of other ethnic groups out in the cold. Anna Need and David Goldstein point out that of the 370 or so GWAS studies conducted up to 2009, 85 percent were on European populations. Only one out of the 370 was conducted on African or Hispanic populations, and a mere 10 percent on Asian populations.

Of course, genetic scientists wouldn't be making the argument for using race in association studies unless some signal of stratification existed and unless there were some instances in which the approach actually told them something. One of the major arguments for using the stratification approach is that it makes the piles of hay in which the genes/needles are hidden smaller and more manageable. Such is the rationale for using ethnic groups like the Icelandic people in case-control studies. By using a cohesive group of people stratified by ethnicity or geographic isolation, the expected amount of variation is decreased and the stack of hay is made smaller. And the statistical methods that are used become stronger when "noise" is reduced by stratification. But whether this advantage is real depends on two things. First, the stratification really does need to reduce variation to make the analysis work better. We have already seen how the isolation of Icelanders might not have been as complete as researchers originally thought, so that the reduction of variation in this population might not be as extreme and therefore not as helpful as had originally been believed. In the case of Iceland the pile of hay actually grows rather than shrinks! The second assumption that needs to be accurate if the study is to be relied upon is that there is little or no substructuring of the populations concerned. And in this regard populations with deep evolutionary histories, like African ones, are going to be extremely difficult to study using GWAS.

We would also point out that stratifying on the basis of race (whatever definition of race you use) only "kind of" works. As we have seen, in most cases statistical methods need to be pushed to their limits, and novel statistical approaches applied, to eke out positive results when using race. There is, of course, a good reason why using race "kind of" works for the HapMap project. To the extent that it works, it does so because the historical remnants of earlier isolation among human populations have left a weak signal that can, in very limited cases, be useful in detecting the association of disease with genes. We do not deny the existence of that weak signal, but we do question its more general utility. In a similar spirit, the human geneticist David Goldstein has suggested that the HapMap studies done on complex psychiatric disorders such as schizophrenia and bipolar disorder demonstrate that almost no correlations can be made. Thus, when a disorder like Type 2 diabetes is examined, only 2 to 3 percent of familial clustering can be explained using the HapMap approach. In addition, when personalized genomics using next-generation sequencing platforms is added to the fray, the claims made for GWAS case-control approaches as necessary, or even as "the best we have at hand," weaken.

GETTING UP CLOSE AND PERSONAL

What about personalized genomics—the process of whole genome sequencing for individuals? We have already discussed the use of personalized genomics with respect to Miller syndrome and Charcot-Marie-Tooth syndrome, and these two studies show the power of applying personalized genomes to disease study. Actually, the feasibility of doing human genomics with individualized genomes is, above all, what makes the use of race in HapMap so dubious for us. Individualized genomics is a viable alternative approach that is already being developed, and it will most likely be the way we regularly do things within a decade, as the "$1,000 genome" is developed. Nonetheless, it is still a distant goal to have any one study sequence hundreds or thousands of genomes.

Today a full genome costs around $20,000, which is much less than a few years ago but still means that sequencing one hundred genomes would cost two million dollars. But while recent sequencing projects have targeted individuals, who were sequenced at substantial cost, these studies have demonstrated two important things. First, it will soon be possible to sequence an individual's genome not only rapidly but inexpensively. The technology is exploding with innovation and invention. As soon as an approach is dubbed "next generation," the real next generation approach appears, and it is usually at least an order of magnitude faster than its predecessor. We have already noted the increase in sequencing power that has been seen over the last few decades, and it is clear that the technology will advance steadily in coming years. The second thing demonstrated by the first dozen or so human genomes sequenced is that the comparison of whole genomes, on an individual level, will indeed be useful in medicine and in the discovery of disease-related genes. By doing nothing more than comparing the two famous genomes of James Watson and Craig Venter, we can make many points about individualized medicine. For, as P. C. Ng and colleagues stated, "Their genetic differences underscore the importance of personalized genomics over a race-based approach to medicine. To attain truly personalized medicine, the scientific community must aim to elucidate the genetic and environmental factors that contribute to drug reactions and not be satisfied with a simple race-based approach." In other words, to the very people who sequenced these whole genomes, the race-based approach is "simple," basically meaning "simplistic." A wonderful example of the precision of the genome-based approach is provided by the genes called Cytochrome P450s, or CYPs for short, of which each of us has several copies. These genes are involved in many functions, of which one is in the metabolism of pharmaceuticals that are commonly used in treating patients with various disorders. It is an important matter for individualized medicine to be able to predict if a particular drug will be useful in treating someone, so knowing about these genes is important in determining the efficacy of pharmaceutical metabolism. The race-based approach would suggest that both Watson and Venter would respond to the drugs that members of their supposed racial group are best-matched to. Both men self-identify as Caucasians, and on this assumption both men would be treated with the same drug. However, when the whole genome sequences of the two men are examined, it is evident that Venter is homozygous for an allele called CYP2D6. This is a kick-ass CYP allele that makes him an extensive metabolizer. Watson, on the other hand, is homozygous for an allele called "star10" or *10. This allele in Watson's genome has a lower activity for metabolizing pharmaceuticals. Speaking of kick-ass alleles for metabolizing pharmaceuticals, this time recreational ones, researchers have recently reported on several genetic variants in venerable rock and roller Ozzie Osbourne's genome. Mr. Osbourne describes his metabolic proclivity for pharmaceuticals as follows: "Given the swimming pools of booze I've guzzled over the years—not to mention all of the cocaine, morphine, sleeping pills, cough syrup, LSD, Rohypnol . . . you name it—there's really no plausible medical reason why I should still be alive." The scientists who examined his genome discovered several rare variants in Ozzie's genome that could be the reason for his metabolic weirdness. Nathan Pearson of Knome (the company that did the DNA sequencing) explained that based on the genetics of his genome Mr. Osbourne is about three times more likely to react to marijuana with hallucinations and has "an increased predisposition for alcohol dependence of something like six times higher." Personalized genomics meets Black Sabbath.

Let's return to Watson's *10 allele, which raises a tricky concept. This *10 allele is not found in high frequency in Caucasian populations, and indeed it more closely resembles alleles that are found in East Asian populations. In fact, as Ng and colleagues point out, "Watson's genotype predicts that he is likely to differentially metabolize drugs such as antidepressants, antipsychotics, and the cancer drug tamoxifen." In other words, he would respond to these drugs much as an East Asian might. On the flip side, we are sure there are some people who identify themselves as of East Asian ethnicity who would respond to drugs much as Venter would.

When we compare personalized genomes with GWAS approaches, an important point needs to be made with respect to their application to different ethnic groups. We have already seen from Need and Goldstein that there is a huge disparity in number between studies applied to European populations versus other ethnic groups. However, these authors also point out that such a disparity is also a distinct possibility for personalized genomics. They point out that GWAS studies "flirted" with disaster, which was only averted by their lack of success. Had they been successful, and effective treatments had arisen from the GWAS analysis, then a huge disparity would have emerged, with European populations benefiting eight to ten times more from the approach than other ethnic groups. Need and Goldstein further say that "whole genome sequencing efforts will not be 'blessed' by such failures (such as GWAS)." In other words, whole genome approaches will be much more successful, and hence we will then have to face the problem of a huge inequity in genomics research with respect to ethnicity. Need and Goldstein warn that "a dedicated effort to ensure even application of genetic research across major human population groups is a key priority to avoid large gaps in our knowledge of the human genome." Note that the focus of the warning is not on the genome of this or that group but on the "human genome." The human genome is, after all, the sum collection of genomes from our species across the globe.

HAS THE USE OF RACE IN HUMAN DISEASE TREATMENT WORKED?

Even though next-generation genome technologies are being developed, and will be available eventually as medical tools, we can still ask if there are certain circumstances in which racial categorization has helped, or would help, in the control of disease. Examples concern the development and distribution of pharmaceuticals targeted at so-called racial groups. While the history of using race in developing treatments is fraught with horrible events such as the syphilis "treatment" of African American men (the "Tuskegee experiment"), proponents of the race-based pharmaceutical approach cite the large number of diseases that appear to have a "racial" component as justification for a race-based approach. Table 6 lists several diseases reported to have high risks and low risks in different so-called "ethnic groups."

Stratification along the lines reflected in this table is highly arbitrary and makes little sense with respect to any unified idea of what races are. In some cases, the racial lines are drawn based on the geographic idea of three races; in others they are drawn more specifically (as in multiple sclerosis); and in still others, the use of hugely admixed "racial" groups (as in osteoporosis) is obvious. It is this inconsistency that has led a famous geneticist, the former head of the NHGRI (National Human Genomics Research Institute at the National Institutes of Health) and currently head of NIH, to suggest, entirely correctly, that "'race' and 'ethnicity' are poorly defined terms that serve as flawed surrogates for multiple environmental and genetic factors in disease causation, including ancestral geographic origins, socioeconomic status, education, and access to health care."

In addition to the application of "race" in studies of this kind, the few cases of the development of "race-based pharmaceuticals" to combat various "race-based diseases" have largely failed in practice. BiDil is perhaps the best example of a disappointing race-targeted drug. BiDil was developed as an anti–heart attack agent and was found by the company Nitromed to have better efficacy among African Americans. It was specifically conceived as a race-based drug because "at this point the technology and resources do not exist to scan efficiently every individual's genetic profile" to determine its efficacy. It was approved by the FDA for general clinical usage in 2005, as a result of a large clinical trial using the African American Heart Failure Trial (A-HeFT), a study co-sponsored by NitroMed and the Association of Black Cardiologists, and it has been available on the market for several years. BiDil's efficacy was established by reliance upon clinical tests on individuals who self-identified as having African ancestry. This was something of a problem, because it stacked the deck for efficacy. Another problem with the drug and the statements about its effectiveness is that while it is not universally efficient on all African Americans, it is actually quite efficaceous in many individuals who self-identify as Caucasians. And although BiDil is still available on the market, it has not been very successful financially (possibly in part because it was overpriced), and NitroMed has admitted the lack of a market for the drug.

Having hung its success on BiDil, Nitromed is now exploring strategic options, including sale of the company. So why did the market evaporate? Or, more accurately,

TABLE 6. Diseases reported to be correlated with geography

Disease	High Risk	Low Risk
Obesity	African women, Native Americans, South Asians, Pacific Islanders, Aboriginal Australians	Europeans
Non–insulin dependent diabetes	South Asians, West Africans, Peninsular Arabs, Pacific Islanders, Native Americans	Europeans
Hypertension	African Americans, West Africans	Europeans
Coronary heart disease	South Asians	West Africans
End-stage renal disease	Native Americans and Africans	Europeans
Dementia	Europeans	African Americans, Hispanic Americans
Systemic lupus erythematosus	West Africans, Native Americans	Europeans
Skin cancer	Europeans	
Lung cancer	Africans, European Americans (Caucasians)	Chinese, Japanese
Prostate cancer	Africans and African Americans	
Multiple sclerosis	Europeans, African Americans, Turkmens, Native Siberians, New Zealand Maoris	Chinese, Japanese
Osteoporosis	European Americans	African Americans

why was the market not as receptive to the drug as Nitromed might have thought? We suggest that, in order for any drug to have a viable demand from a targeted market, that market has to exist in the first place. And the general public is not as naïve as some marketers might think. As Jonathan Kahn suggests, "medical researchers may say they are using race as a surrogate to target biology in drug development, but corporations are using biology as a surrogate to target race in drug marketing." Such market-driven targeting of race in developing treatments for people is more than likely the reason for the lack of resounding success of BiDil.

So when we examine the oft-used justification for race in medicine and the study of disease, we see that all three of the major questions we raised about the use of race in these contexts clearly point to a *lack* of utility. And while many smart scientists are still suggesting that this approach works, it is highly doubtful that it has been as useful as it has been touted to be or ever will be. There are many reasons why science doesn't behave the way the proponents of using race as a criterion want it to, and there are certainly better and more efficient ways of studying disease and developing pharmaceuticals for treatment now being developed. Specifically, with the $1,000 genome on the horizon, we will soon have the ultimate tool for individualized medicine. Some feel that the use of individualized genomes will also have its problems. We do not doubt this. However, we suggest that the use of race in personalized medicine will not be one of them. In the meantime, more precise genetic approaches, such as the trio and cohort techniques, are conceptually and theoretically superior to race-based approaches to medical genetics.

IF HE WERE ALIVE TODAY, WOULD DARWIN PAY FOR A GENETIC ANCESTRY TEST?

While writing this book, we canvassed the internet for commercial DNA ancestry labs and found at least twenty of them. These outfits are genetics companies that take mouth swabs or blood samples from their customers, isolate DNA from the cells contained in the swabs, and sequence several genes or analyze thousands of markers using micro-array methods to do two things. They first look at the customer's genome for anomalous (or normal) genetic variation in a large suite of genes. Most of the companies doing this are in the business of analyzing a panel of "disease" genes for the interested consumer. Second, they claim to be able to establish the ancestry of the customer's genome. In this context there is a maternal and (if male) a paternal lineage that can be pinpointed using DNA sequence technology, and some companies also claim that, using autosomal (non-sex-chromosome) markers, an estimate of a person's admixture can be determined. While some companies focus on one or the other of these two products, some companies also offer packages for both. With respect to ancestry, these companies make claims as enticing as those in any effective marketing campaign. Below are four of many examples we found:

Genetic Ancestry Analysis:
>What's Your Tribe?
>Compare your DNA to the largest global data base of over 900 ethnic populations and 36 world regions.

DNA Ancestry Project:
>Trace your ancestral origins—
>Learn more about your ancestors, where they came from, and their ethnic origins.

Learn why ancestry is the answer.
>Discover your maternal roots.
>Discover your paternal roots.

Compare your DNA to populations around the world.
>We'll help you find the homeland of your ancestors.

The marketing ploy used by these companies is to push the "discovery" of your ancestry as an important aspect of your humanity. What tribe do you belong to? Find out where your homeland is! Learn more about your ancestors! These are pretty fantastic claims, and some companies concerned succeeded initially, although recently some have been failing. Still, when you throw into the mix the efforts of other, well respected outfits such as the venerable scientific institution the National Geographic Society and its "Genographic project," the general air of excitement over human ancestry becomes evident. But what should also be evident is that the central thrust here is geographical origin, rather than an analysis of the human condition itself.

In which case, perhaps we would do better to proceed with caution. There are two reasons why we feel it important to examine the premise that a person's ancestry can be determined first by examining closely the two most popular molecular markers—mtDNA and Y chromosomal DNA—used for this task. As we have already said

several times, when we use these tools we are not reconstructing the historical relationships of people. Rather, we are reconstructing the relationships of tiny little
chunks of people's genomes. Second, practitioners of both of these aspects of human
DNA research have long made tracing genealogy more or less synonymous with tracing race. In the words of the journalist Sally Lehrman, "genetic researchers have
begun . . . replacing 'race' with 'ancestry.'" Many currently active private genetics
companies suggest they can directly trace a customer's ancestry; so saying that your
mtDNA has a certain haplotype, and hence a certain relationship to an ancestor,
involves equating your haplotype with your ethnicity or "race." Such conflation of
ancestry and race can be both scientifically inaccurate and socially undesirable, as
we hope to show in the next sections. In a similar spirit, Deborah Bolnick and a
group of co-authors from a surprising range of specialties recently warned in the august pages of *Science* magazine that "we must weigh the risks and benefits of genetic
ancestry testing." In addition to the well-established maternal and paternal markers,
some companies use those "ancestry-informative markers" (AIMs) that reside mostly
in the autosomal component of the human genome, and we will have much to say
about this manner of proceeding as well.

If Darwin were alive today, we wonder whether he would be amused or annoyed
by the proliferation of genetic ancestry companies that promise to determine the ancestry of people in the consumer audience. Whether he would or not, there are two
aspects of ancestry raised by his own pedigree that we think would intrigue him greatly.
We thus traced Darwin's mitochondrial lineage and his Y chromosomal lineage, superimposed on his pedigree over the ten generations available for his ancestors (back five
generations) and his descendants (forward five generations). The family tree we used
comes from the "Pedigree of the Darwin-Wedgwood-Galton Family" first presented at
the American Museum of Natural History at the 3rd International Eugenics Conference in 1932 and from detailed pedigrees subsequently developed by R. B. Freeman.

What this allows us to do is get a picture of the shape of the two trees for these
two markers in Darwin's family. When the Y chromosome tree is drawn by tracing lines
through Darwin's father, his father's father, his father's father's father, his father's father's
father's father and his father's father's father's father's father, as well as to his sons, his
sons' sons, his sons' sons' sons, his sons' sons' sons' sons, and his sons' sons' sons' sons'
sons, the tree has several branches and is quite complicated. On the other hand, Darwin's mtDNA tree has many fewer branches and divergence points, because few of his

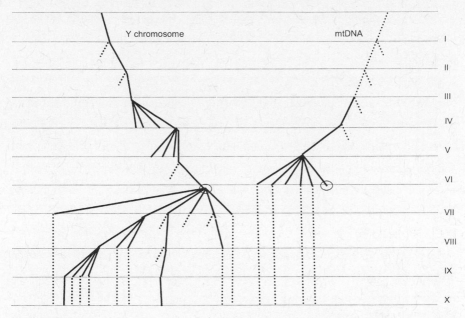

FIGURE 9. *Tracings of Charles Darwin's Y chromosome (left) and mtDNA (right) from the "Pedigree of the Darwin-Wedgewood-Galton Family" first presented at the infamous 3rd International Eugenics Conference in 1932. Solid lines represent actual ancestor-descendant events, while vertical dotted lines are "hypothetical" descendants traced out to facilitate drawing of the trees. Slanted dotted lines represent potential branch-points, to demonstrate generational branches. The black circle represents Darwin's position in the network. Horzontal dotted lines represent the generation boundaries in Darwin's pedigree, and the Roman numerals represent generation numbers, starting with Darwin's great-great-great-grandparents in generation I.*

female ancestors produced female offspring, and few of his female siblings (who would also have had his mtDNA) did either. Darwin would almost certainly have noticed that the trees for his Y chromosome and for his mtDNA are rather different in their topologies and hence in the patterns of divergence that they convey. Given that the only figure in *On the Origin of Species* (as you will recall, a genealogy) testifies that Darwin had a very sophisticated understanding of genealogy and history, we suspect that he would have recognized immediately that, for these two genetic markers, he and his ancestors and descendants have two very different histories.

We also suspect that Darwin would note from this exercise that he is an mtDNA dead end. He would see straight away that none of his descendants have his mother's (and hence his) mitochondrial DNA, not only because of course he doesn't pass his mtDNA to his offspring, but also because his four sisters never reproduced. He would also notice that his Y chromosome DNA has fared only slightly better, since it can be found today, as far as we could tell, in only two of his male descendants (a father and son living in Australia).

The two crucial points that all this illustrates are frequently lost on the consumer public when they employ DNA companies to trace their ancestries, whether mtDNA, Y chromosomes, or autosomal SNP analyses are analyzed. First, when a company traces someone's ancestry, it is tracing the ancestry of a *marker* (a mtDNA marker or a Y chromosomal marker or an autosomal SNP marker), not of an *individual*. That individual may have a specific genetic ancestry somehow reflected in his or her mtDNA lineage, but he or she is also a mosaic of the many thousands of genes that reside in the human genome. In some cases, markers are used that are distributed more broadly across the genome, but exactly the same can be said for these other, autosomal, markers. Second, in some instances written or even oral knowledge of ancestry can be more predictive than the actual molecular tracings that these two particular markers accomplish. It should be no surprise to readers with some knowledge of genetics that males we know to be in Darwin's lineage—with the surnames Keynes, Barlow, Cornford, and Chapman, among others—have mtDNA and Y chromosomal markers that do not link them back to the most famous of their ancestors—Charles Darwin.

Other ancestry-determining approaches, using those AIMs we just mentioned (and discuss in detail below), have also been used in similar contexts. A recent PBS show called *Faces of America* focused on this approach as a means of telling several celebrity guests about their genetic background. Each guest (except for one who passed on the exercise) was handed a pie chart revealing what percentage of him or her was African, Caucasian, or Asian. Guests included Stephen Colbert (100 percent Caucasian), YoYo Ma (100 percent Asian), and Eva Longoria (70 percent Caucasian, 27 percent Asian, and 3 percent African).

If one makes the assumption that the AIM markers truly indicate "race" then, okay, Steven Colbert is 100 percent Caucasian, and, as Mr. Colbert himself says, he constitutes "the inescapable black hole of white people."

However, there are problems with the interpretation of these markers. Charles Rotimi, a researcher at the National Institutes of Health, has stated that "the nature or appearance of genetic clustering (grouping) of people is a function of how populations are sampled, of how criteria for boundaries between clusters are set, and of the level of resolution used." In other words, these ancestry markers can mislead us if we are not careful. We would also add that these markers are for the most part cherry-picked. Would another set of markers from Mr. Colbert's genome, for instance, yield a different pie chart? Would this mean he somehow has a multiple existence? Let's look at another way to interpret the ancestry testing using these large numbers of SNPs.

We begin by asking what precisely AIMs are. The simplest explanation is that they are essentially the same kind of markers, discussed in chapter 3, that have been used to type people to country in Europe, and hence they have the same problems that we discussed in that chapter. More specifically, they are examined by using the high throughput approach called a beadchip (the details of which are amazing but not necessary for us to understand the genetics discussed here). Beadchips can determine the DNA sequence of SNPs at large numbers of places in the human genome. For instance, two of the more prominent personalized genomics companies out there, one called 23andme and the other called deCODE, use genome-level tests with over five hundred thousand SNPs and one million SNPs respectively. These companies use what is called an ancestry painting approach to convey to their consumers the ancestry patterns these tests produce. Basically, the AIM SNPs from an analysis are classified as belonging to this or that race, and because their locations on specific chromosomes are known, those regions of the chromosome get "painted" with a color that corresponds to a particular racial group, say blue for Europeans, green for Africans, and red for Asians. The result is a very colorful diagram of the twenty-three chromosomes of a human being that claims to show his or her ancestry. In addition, these companies claim to be able to be even more precise for people of European ancestry, by pinning down their ancestry to specific geographic regions within Europe.

But how exactly are these markers determined, and how many of them are there? If all half a million to a million of these markers were AIM informative (were indicative of race) our book would have been very short indeed. But it turns out that only a very small proportion of the SNPs in these tests are AIM informative. Because of the proprietary nature of the AIMs used by these companies, we are stuck with guessing at how many and which they use. A lower-end guess we can make comes from the work of Michael Seldin and his colleagues who have established what they believe is a panel of 128 ancestry-informative markers for determining continental origin (their most recent work claims to have whittled the number necessary for accuracy down to 93). In another study, Michael Smith and colleagues claim to have

found 3,011 markers that are AIM informative. So in the following number crunch-
ing to estimate the number of AIM-informative SNPs used, we will take the higher
end estimate of 3,011 as settled on by Smith and colleagues. The math is pretty easy.
Only 3 of the markers are informative (10^3 out of 10^6 are uninformative), meaning
0.997 or 99.7 percent of the markers are AIM uninformative. We still have a problem because what "uninformative" means is debatable. But we *can* say that there are
two kinds of uninformative markers. First there are markers that are so polymorphic
that they do not have the information to tell what so-called race they are indicative
of. These polymorphic markers probably derive from the ancestral pool of human
genomes that gave rise to the great diversity of humans we see today. In other words,
they are most likely African. The second category of uninformative markers are those
that are found in only a few individuals. These markers are most likely what we can
call derived, and since they are unique to only a small number of humans they cannot
be indicative of any so-called race.

When we look at the several human genomes that have been fully sequenced
so far, it is somewhat difficult to tease apart these ancestral and derived markers. But
a quick back-of-the-envelope calculation reveals that about one-third of the markers
are found in more than one so-called racial group (see discussion of the genomes of
Bishop Tutu, !Gubi, G/aq'o, D#kgao, !Ai, and others earlier in this chapter). The
other two-thirds of the markers are unique to individuals. We should note that this
way of calculating ancestral markers versus derived markers will underestimate the
number of ancestral markers because of the small sample of fully sequenced genomes
to date. But this exercise does reveal something very telling about what the AIM
markers leave out. If 300,000 (roughly one-third of 1,000,000 markers total) are ancestral or African to begin with, then 30 percent of all one million of the markers
are African to begin with. If we look only at those markers that are informative (the
300,000 African markers plus the 3,000 or so that are AIM informative, then only
1 percent are AIM informative to a specific so-called race, and 99 percent are African (3,000 versus 300,000).

To us this means that Steven Colbert's pie chart changes from 100 percent Caucasian to 99 percent African and 1 percent Caucasian, YoYo Ma's changes from 100 percent Asian to 99 percent African and 1 percent Asian, and Eva Longoria's changes
from 70 percent Caucasian, 27 percent Asian, and 3 percent African to 99.03 percent

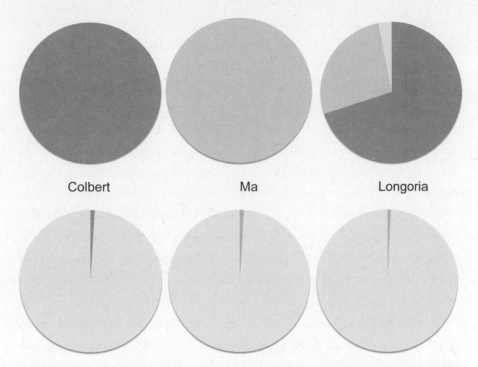

FIGURE 10. *"Ancestry" pie charts for three celebrities as presented to them on the PBS-TV program* Faces of America *(top row). Bottom row shows the pie charts as recalculated by the authors using estimates of variation from the entire genome. Dark gray = "European"; medium gray = "Asian"; light gray = "African."*

African, 0.7 percent Caucasian, and 0.27 percent Asian. These pie charts would all look almost identical (see figure 10).

The broad differences the ancestry testing companies point to from the AIM-informative markers become the major focal point of comparison for their consumers. People see the *differences* as significant and focus on those differences as being biologically meaningful. But when we include all markers, the new focal point shows how remarkably alike all the pie charts look. Instead of Stephen Colbert standing out as "the inescapable black hole of white people" he becomes pretty much as African as the rest of us. This is not to deny his recent European ancestry but rather to recognize his slightly more distant, and what we consider stronger, connection through ancestry with the rest of mankind that undoubtedly came ultimately from Africa. In essence, the set of markers chosen by ancestry-testing companies has been artificially reified by the companies and scientists concerned, as a tool for demonstrating something—"biological race"—that actually does not exist.

While some companies are very clear in their online tutorials about the limitations of the product they offer, we wonder how many people in the general consumer public are really aware of the paradoxes and problems when they shell out for their DNA genealogies or for their admixture pie charts? We also suspect that many cases of the kind of lineage-sorting that we observe over a mere ten generations in Darwin's pedigree might also occur over longer evolutionary periods. In fact, people are routinely shocked by the results of their DNA tests. Henry Louis Gates, the distinguished African American scholar at Harvard University and the host of the PBS show mentioned above, was reportedly taken aback when his DNA tests revealed a large component of European ancestry in his genome (perhaps he shouldn't have been so surprised: one recent study using AIMs suggested that the European contribution to the ancestry of a sample of residents of Columbia, South Carolina, who self-identified as "black" ranged up to over 50 percent). Still, it took Gates, one of the giants of the literature detailing the African experience in modern society, only a short time to distinguish the central cultural component of his ethnicity from any unexpected genetic component of his ancestry. The cultural part of his African heritage, which is very clear to anyone who has read his writing, and who has acquired a feel for his interpretation of what it means to be an African American, vastly overrides any genetic anomaly that these tests might reveal. In addition, if one watches the *Faces of America* show closely, it becomes obvious that what had the most moving and informative impact on Dr. Gates's twelve guests was the written and oral information on their backgrounds that they were presented with. The genetic information was basically a convenient, colorful, and sexy add-on. Overall, though, most of the guests on this important PBS offering clearly recognized intuitively that, in a very real sense, we are all who we *think* we are.

None of this is to deny the very real pull that ancestry testing seems to exert on people. As human beings we are intensely interested in where we came from, both as individuals and as a species. Ancestry testing helps us to satisfy this thirst to know about ourselves, and to this extent it satisfies a very basic need. Still, we should remember the underlying biological message: for all the cultural and genealogical differences that help mold our identities, in a biological sense it is our similarities that are very much more meaningful.

COURTROOM DRAMA

Most readers will be familiar with the many television series that use modern forensic approaches to examine crime scenes and provide fodder for TV courtroom dramas. While these stories might seem farfetched to the critical eye, to the naïve viewer they appear within the realm of reality. Most such stories exploit common mythology about DNA sequencing, but they are not clear about the approaches used, and there is no indication of the controversial nature of some of the techniques. This having been said, it is nonetheless true that individual identification/exclusion is an important and well-tested approach in forensics. And it relies on the large amount of variability that exists in all human populations.

Remember, there are fifteen million SNPs in the human genome discovered so far. This doesn't mean there are fifteen million different kinds of human genomes in existence on the planet. If this were the case, the identification of one individual out of the six billion now alive on the planet would be impossible using just these SNPs and allied approaches. But human genomes recombine during the production of gametes, and this ensures that virtually every human being on the planet has a different set of SNPs or other kinds of markers: identical twins are the only potential exception (a very small number of differences do exist between twins).

Most forensics is done with microsatellites, those parts of the genome that vary in size because of variation in the number of small two, three, and four base pair repeats we discussed earlier. The mathematics of probability for the matching of different microsatellite markers with the patterns found at the same loci of a suspect is pretty simple, but in practice more than one or two (indeed, typically thirteen) markers need to be examined to exclude someone as a match for a particular profile. This is because if a person matches to one, or two, or even three microsatellite markers taken from a forensic sample, there is still a very high probability that the sample did not originate from that person. The issue thus boils down to one of probability. Matches at three marker loci could mean that there is only a one-in-one-hundred-thousand chance that the forensic sample is not that person's. But given that there are six billion people on the planet, one in one hundred thousand is not good enough. In fact, if this number were produced for a forensic study in New York City (where, if you follow TV, most of the mayhem in the country appears to be committed), then (since there are about eight million people in NYC), there might be about eighty people with those three markers walking the city's streets.

On the other hand, if there are matches at a grand majority of the marker loci (like blood from O.J.'s socks matched to Nicole Simpson's blood from the murder site) there would be a one in ten-to-twenty-billion chance that the blood examined was not the victim's. Even if you included the possibility that every single human on the planet could be the source of the sample in the notorious Simpson trial, the results presented mean that there would have to be twenty billion people on our planet for us even to entertain the possibility that there is another person out there with the same genetic markers.

There are many things that need to be in place in order for the forensic approach to work. First, the DNA obtained from a crime scene sample needs to be uncompromised. This means that any sort of contamination or tinkering with the sample would invalidate its use. This is why the CSI (CBS) guys always rush to a crime scene and make sure it is secure, and why, in his capacity as a blood splatter analyst, Dexter (Showtime) can get away with murder by covertly substituting crime scene samples. If all goes well at the crime scene, then to catch a criminal the database has to have the profile for the microsatellites the forensics lab uses. No database, no match. For instance, neither of the authors of this book has been typed for any kind of DNA, so our profiles don't exist in any database of any sort. This would mean that if we committed a crime and left skin, blood, hair, or any number of body parts behind, we could not be detected by the DNA forensics.

Of course, this doesn't mean that the two of us would go scot-free on the basis of DNA, because the first thing any good district attorney would do if DNA were found at a crime scene would be to subpoena any possible suspect for a DNA sample. In that case, we both would have to give up some mouth skin cells, and (if guilty) away to prison we would go. The DNA Identification Act was passed by the U.S. Congress in 1994, and all fifty states now have laws on the books that require certain kinds of offenders to have their DNA profiles on file. The database that stores these profiles is called the Combined DNA Index System (CODIS). Currently there are five million offenders who have their DNA profiles on file in CODIS, and there are another two hundred thousand or so profiles from crime scenes that have not been identified with particular offenders. In addition, the military and some corporations obtain DNA profiles of their members.

By the way, this technology works in the other direction too, as the diligent work of Barry Scheck and the Innocence Project has demonstrated over the past couple of decades. The Innocence Project takes capital cases and reapplies DNA sequence marker technology to them. To date, 249 individuals on death row have been exonerated by the Innocence Project's ability to exclude these wrongly convicted death row inmates from commission of the crimes they were convicted of.

So far, the requirements for the application of the techniques (having a database and having uncompromised crime scenes) theoretically do not involve race. However, since a disproportionate number of minority males are in prison, there will be a disproportionate number of men from these cultural ethnic groups in the database. And race has been creeping into the picture in another way. Because the CODIS database only has five million entries in it, there is a high probability that DNA from a crime scene will go without an identification hit in CODIS. Without the match the initial information is not useful, unless the criminal is caught, tested, and subsequently matched to the DNA from the crime scene.

For the naïve, it does not appear a big leap to reason that the DNA profile can be quite accurately matched to a "racial" group and that this information could then help to catch the criminal. In fact, the very same reasoning is being used in Europe to test people for their nation of origin for immigration purposes, in an initiative called the "Human Provenance Pilot Project." We have already explained at some length why we feel this approach is unworkable. As we saw, it is very dubious that, as claimed, people from Europe can be identified to within four hundred kilometers of their "origin." Indeed, there are so many problems with this approach that one geneticist, Mark Thomas, was quoted in the journal *Science* as saying that it is "little better than genetic astrology." To us, it would be infinitely better to use DNA profiling to identify individuals directly, as will soon be possible. In other words, the solution is the same as in race and medicine—individualize it.

EPILOGUE

AS THIS BOOK nears its finish, we'd like to draw the reader's attention to an interesting fact that we feel is important to bear in mind when we think about how we, as a species, view our own biology, and why we have probably had much more than enough already of "race" as a biological concept. The development of both evolutionary theory and high-throughput genetic technology has happened during a uniquely busy phase in the biology and history of *Homo sapiens*. We have recently developed a scientific framework (evolution and genomics) for understanding the organization of our natural world, right on the heels of the historical biological reintegration we discussed earlier in this book. While already deeply eroded, the genetic signals of ancestry and group membership are still discernible at this point in human history, as testified by the occasional correlation (mostly weak) of the genetic and physical appearances of people within geographic or ethnic groupings. But think what the situation would be if this evolutionary theory, and these highly precise genetic technologies, had not been developed for another five hundred or a thousand years, even as present demographic trends continued. If our species makes it that far into the future, it is highly likely that by then the genetic and visible signals for both ancestry and group-relatedness will have been even more severely attenuated. For it is clear that many of the physical differences that exist amongst populations *are* slowly fading. Recently Nina Jablonski, an expert on human skin color, suggested in an interview that in the near future "there will continue to be isolated areas with people at the extremes, but in major urban environments there will be much more mixing and eventually matching of color." Our genes will behave the same way, and ethnic and linguistic differences will also tend to blur in an increasingly affluent and globalized world, although they do act to slow down processes of biological blending that in another polytypic mammal species would continue more freely under similar circumstances.

Even the fine-scale genetic signals discussed earlier in this book that occasionally correlate with disease and other phenotypes may not exist in another millennium or two. And this harkens back yet again to the insight, now many decades old, that there is more within-group variation among humans than among-group variation. As the reintegration process continues, with greater and greater intensity, the visible gaps between populations will begin to vanish. At the same time, the levels of variation *within* populations will only increase, leaving us with an even greater "Lewontin effect" in human populations. We suppose that there are some good things to say about the fact that we happen to have caught ourselves at this particular point in our population biology. If the appropriate population signals are there, we should certainly use them to study disease, but we should not confuse this steadily dissolving phenomenon with "race." As scientists, we have to resist several temptations when applying our new technologies to our understanding of evolution and human ethnicity. One of these temptations involves making the leap from using stratification as an algorithmic tool in genetics to recognizing the resulting stratifications as "races." Another is to use genes willy-nilly as markers for the evolution of individuals, without considering the biology of those genes, and to make the mistake of equating the history of gene X with that of individual A.

Perhaps, though, the most important thing of all to bear in mind when we are trying to place the race issue in perspective is that our species *Homo sapiens* is of very recent origin. Most species on this Earth have been around a lot longer than our mere two hundred thousand years. And this means, of course, that all of the differences within our species that we perceive as racial are even younger. Indeed, diverse lines of evidence converge on the conclusion that most of the variation we perceive has emerged—and has commenced to blend again—within the last fifty thousand years or thereabouts. So, in a very real sense, our population differences are highly superficial—and what's more, as variations within a single species, they are also ephemeral. As we've seen, even though broad-brush characterizations of racial groups can be made, it is impossible to draw hard and fast lines either among or within the larger regional populations that exist today. And it is a manifest fact that although *Homo sapiens*, the most cosmopolitan of all mammal species, has ramified just as all successful species are wont to do, it has witnessed no speciation. We all remain part of one big if only intermittently happy family (breeding species) that has predictably diversified in the short time since its origin.

During most of the brief period of our species' existence, human populations were thinly scattered across a highly changeable globe. Isolated tiny populations plus unstable environments equals the most propitious conditions for local fixation of heritable novelties; and as a result the "racial" differences we see today emerged in far-flung parts of the Old World. This was the interlude during which at various points the populations we know today broadly as San, Africans, Asians, Australian Aborigines, and Europeans acquired their recognizable local features. But since the end of the Ice Ages, the rules of the evolutionary game have fundamentally changed, at least as they apply to humans—entirely as a result of those cognitive innovations that propelled our species' initial rapid spread.

Around ten or twelve thousand years ago, as climates warmed up, humans began to take up settled lifestyles in response to changing environmental conditions. Once again, we see the typical but largely unprecedented modern human response to environmental challenge: the invention of new technologies to cope with new conditions, rather than adaptation of old tool kits to new uses. Over a remarkably short span of time, and independently in several areas of the world, the immemorial itinerant hunting and gathering lifestyle was replaced: first, by semi-permanent occupation of favored sites, and then by the formation of permanent villages and the domestication of animals and crops. With agriculture came a pressing need for hands to till the fields and herd the sheep and goats, or whatever animals it might have been that were domesticated in a particular locality; and population soared, as the settled lifestyle made it possible for mothers to cope with multiple children at once.

Thus the human population was set on a trajectory toward the extraordinary and still-increasing density we see in the world today. And, as agricultural and herding populations grew (simultaneously from several centers, ultimately squeezing and marginalizing the itinerant San and Aborigines and other hunter-gatherers), new evolutionary rules kicked in. Previously isolated populations expanded and came into contact. Doubtless they frequently also came into conflict—but, as members of the same species, they nonetheless readily exchanged genes. From a story of local differentiation, the history of the rapidly multiplying human species became one of reintegration, blurring the physical distinctions (most of them almost certainly functionally neutral) that had been developed in isolation. Boundaries, such as they may have been, began to break down, albeit in a process slowed by cultural factors.

It is this ongoing, and indeed accelerating, process of reintegration that explains why it's a hopeless and counterproductive task to recognize and categorize discrete "races" or subspecies among *Homo sapiens* today. Indeed, to attempt to do so, as our government still often insists upon doing, militates against the natural biological trend within our densely packed and increasingly mobile species. Ask Ian's stepdaughter, who is fed up with being incessantly asked to classify herself—and, worse, to do so by choosing between fixed, unreal, and increasingly outmoded categories. Or consult Rob's daughters who, when asked to describe people by their skin color, do so using eight or nine shades. We have seen that the dynamic processes at work on human variation defy static classification; and, even when it is viewed as a "snapshot in time," what we observe around us today defies satisfactory categorization. Our innate urge to classify things, including ourselves, serves merely to create racial labels that, in the words of the anthropologist John Relethford, only "crudely and imprecisely describe real variation."

So enough already of race, certainly as a biological unit. Biologically speaking, race is better characterized as a non-problem. It would be amazing if our close-knit but unprecedentedly widespread species had *not* diversified locally under Ice Age conditions. And it would be equally amazing if reintegration were not occurring under the circumstances that obtain today, despite the undoubted cultural barriers that are unique to our species. There is no doubt that the issue of race continues to be a difficult and highly vexed one, on many levels. But to pretend that biology is usefully added to the mix of historical and sociocultural problems surrounding it only makes them needlessly complex, and yet more difficult to resolve.

FOR FURTHER READING

The publications below, organized by chapter, include major resources in the field and all works specifically cited or quoted from in this book.

CHAPTER 1. Race in Western Scientific History

AAPA. 1996. American Association of Physical Anthropologists statement on biological aspects of race. *Amer. Jour. Phys. Anthropol.* 101:569–70.

Agassiz, L. 1850. Diversity of origin of the human races. *Christian Examiner* 49: 110–45.

Andreasen, R. O. 2000. Race: Biological reality or social construct? *Phil. Sci.* 67:S653–S666.

————. 2004. The cladistic race concept: A defense. *Biol. Phil.* 19:425–42.

Balter, M. 2006. Brain man makes waves with claims of recent human evolution. *Science* 314:1871–73.

Blumenbach, J. F. 1775–95. *De generis humane varietate nativa.* Trans. T. Bendyshe, 1865. Elibron Classics.

Brace, L. C., and A. M. F. Montagu. 1965. *Man's evolution: An introduction to physical anthropology.* New York: Macmillan.

Brattain, M. 2007. Race, racism and antiracism: UNESCO and the politics of presenting science to the postwar public. *Amer. Hist. Rev.* 112:1386–1413.

Broca, P. 1862. "Sur l'hybridité de l'homme." In *This is race,* ed. and trans. E. W. Count (1950).

Buffon, G. L. L. Comte de. 1749 et seq. (trans. 1860). "A natural history. General and particular; Containing the history and theory of the earth, etc." In *This is race,* ed. E. W. Count (1950).

Chakraborty, R. 1982. Allocation vs. variation: The issue of genetic differences between human racial groups. *Amer. Naturalist* 120:403–404.

Coon, C. 1962. *The origin of races.* New York: Alfred A. Knopf.

———. 1981. *Adventures and discoveries: The autobiography of Carleton S. Coon, anthropologist and explorer.* Englewood Cliffs, N.J.: Prentice Hall.

Coon, C., and E. E. Hunt. 1965. *The living races of man.* New York: Alfred A. Knopf.

Count, E. W., ed. 1950. *This is race: An anthology selected from the international literature on the races of man.* New York: Henry Schuman.

Cuvier, G. 1817. "Varieties of the human species." In *This is race,* ed. E. W. Count (1950).

Darwin, C. 1859. *On the origin of species by means of natural selection.* London: John Murray.

———. 1871. *The descent of man, and selection in relation to sex.* 2 vols. London: John Murray.

Darwin, E. 1794–96. *Zoonomia, or the laws of organic life.* 2 vols. London: J. Johnson.

Desmond, A., and J. Moore. 2009. *Darwin's sacred cause: How a hatred of slavery shaped Darwin's views on human evolution.* Boston: Houghton Mifflin Harcourt.

Dobzhansky, T. 1937. *Genetics and the origin of species.* New York: Columbia University Press.

———. 1944. On species and races of fossil and living man. *Amer. Jour. Phys. Anthropol.* 2:251–65.

———. 1962. *Mankind evolving.* New York: Columbia University Press.

Edwards, W. F. 1845. De l'influence réciproque des races sur le caractère national. *Mém. Soc. Ethnol. Paris* 2:1–12.

Edwards, A. W. F. 2003. Human genetic diversity: Lewontin's fallacy. *BioEssays* 25: 798–801.

Evans, P. D., S. L. Gilbert, N. Mekel-Bobrov, E. J. Vallender, J. R. Anderson, L. M. Vaez-Azizi, S. A. Tishkoff, R. R. Hudson, and B. T. Lahn. 2005. Microcephalin, a gene regulating brain size, continues to evolve adaptively in humans. *Science* 309:1717–20.

Galton, F. 1908. *Memories of my life.* London: Methuen.

Garn, S. M. 1961. *Human races.* Springfield, Ill.: Charles C. Thomas.

Gobineau, J. A. de. 1853–55. *Essai sur l'inégalité des races humaines. 4 vols.* Trans. as *Essay on the inequality of human races.* New York: Howard Fertig (1999).

Grant, M. 1916. *The passing of the great race*. New York: Scribner.

Graves, J. L., Jr. 2001. *The emperor's new clothes: Biological theories of race at the millennium*. New Brunswick, N.J.: Rutgers University Press.

Gould, S. J. 1981. *The mismeasure of man*. 2d ed., 1994. New York: W. W. Norton.

Haeckel, E. 1868. *Natürlische Schöpfungsgeschichte*. Berlin.

———. *Die welträthsel: Gerimeinverständlische Studien über Menschen*. Bonn: E. Strauss.

Herrnstein, R. J., and C. Murray. 1994. *The bell curve: Intelligence and class structure in American life*. New York: Free Press.

Hooton, E. A. 1931. *Up from the ape*. New York: Macmillan.

Hunt, J. 1864. *The negro's place in nature*. New York: Van Evrie, Horton.

Huxley, T. H. 1863. *Evidence as to man's place in nature*. London: Williams and Norgate.

Huxley, T. H., and A. C. Haddon. 1935. *We Europeans: A survey of "racial" problems*. London: Jonathan Cape.

Jackson, J. P. 2001. "In ways unacademical": The reception of Carleton Coon's *The Origin of Races*. *Jour. Hist. Biol.* 34:247–85.

Jensen, A. 1969. How much can we boost I.Q. and scholastic achievement? *Harvard Educ. Rev.* 39 (1):1–123.

———. 1982. The debunking of scientific fossils and straw persons. *Contemp. Educ. Rev.* 1:121–35.

Kant, I. 1775. *Von den verschiedenen Racen der Menschen*. In *This is race*, ed. and trans. E. W. Count (1950).

Keith, A. 1936. *History from caves: A new theory of the origin of modern races of mankind*. London: British Speleological Association.

Koenig, B. A., S.-J. Lee, and S. Richardson, eds. 2008. *Revisiting race in a genomic age*. New Brunswick, N.J.: Rutgers University Press.

Kosoy R, R. Nassir, C. Tian, P. A. White, L. M. Butler, G. Silva, R. Kittles, M. E. Alarcon-Riquelme, P. K. Gregersen, J. W. Belmont, et al. 2009. Ancestry informative marker sets for determining continental origin and admixture proportions in common populations in America. *Hum. Mutat.* 30:69–78.

Lahn, B. T., and L. Ebenstein. 2009. Let's celebrate human genetic diversity. *Nature* 461:726–28.

Lamarck, J. B. P. A., Chevalier de. 1809. *Philosophie zoologique*. In *This is race*, ed. and trans. E. W. Count (1950).

Lewontin, R. C. 1972. The apportionment of human diversity. *Evol. Biol.* 6:381–98.

———. 2007. Confusions about human races. http://raceandgenomics.ssrc.org.

Linnaeus, C. 1758. *Systema naturae.* 10th ed. Stockholm: Salvius.

Livingstone, F. E. 1962. On the nonexistence of human races. In *The concept of race,* ed. M. F. A. Montagu, 46–60. New York: Free Press.

Marks, J. 2008. Race: Past, present and future. In *Revisiting race in a genomic age,* ed. B. A. Koenig, S.-J. Lee, and S. Richardson, 21–38. New Brunswick, N.J.: Rutgers University Press.

Mayr, E. 1950. Taxonomic categories in fossil hominids. *Cold Spring Harbor Symp. Quant. Biol.* 15:108–18.

Mekel-Bobrov, N., S. L. Gilbert, P. D. Evans, E. J. Vallender, J. R. Anderson, R. R. Hudson, S. A. Tishkoff, and B. T. Lahn. 2005. Ongoing adaptive evolution of ASPM, a brain size determinant in *Homo sapiens. Science* 309:1720–22.

Montagu, M. F. A. 1942. *Man's most dangerous myth: The fallacy of race.* New York: Columbia University Press.

Morton, S. 1839. *Crania Americana.* Philadelphia: John Penington.

Nott, J. C., and G. R. Gliddon. 1854. *Types of mankind: Ethnological researches.* Philadelphia: Lippincott and Grambo.

Painter, N. I. 2010. *The history of white people.* New York: W. W. Norton.

Peyrère, I. de la. 1655. *Prae-Adamitae.* Paris.

Sarich, V., and F. Miele. 2004. *Race: The reality of human differences.* Boulder, Colo.: Westview Press.

Shipman, P. 1994. *The evolution of racism: Human differences and the use and abuse of science.* Cambridge, Mass.: Harvard University Press.

Shockley, W. 1967. A "try simplest cases" approach to the heredity-poverty-crime problem. *Proc. Nat. Acad. Sci. USA* 57:1767–74.

Smith, M. W., J. A. Lautenberger, H. D. Shin, J.-P. Chretien, S. Shrestha, D. A. Gilbert, and S. J. O'Brien. 2001. Markers for mapping by admixture linkage disequilibrium in African American and Hispanic populations. *Am. J. Hum. Genet.* 69: 1080–94.

Spencer, H. 1864. *Principles of Biology.* 2 vols. New York: Appleton.

Stocking, G. W., Jr., ed. 1988. *Bones, bodies, behavior: Essays on biological anthropology.* Madison: University of Wisconsin Press.

Tattersall, I. 2004. Race: Scientific nonproblem, cultural quagmire. *Anat. Rec. (New Anat.)* 278B:23–26.

Topinard, P. 1877. *l'Anthropologie*. Paris: C. Reinwald.

UNESCO. 1950. Statement on race. *Man* 220:138–39.

———. 1951. Statement on race. *Int. Soc. Sci. Bull.* 3:154–58.

Vogt, K. 1864. *Lectures on man: His place in creation, and the history of the earth*. London: Longman.

Wallace, A. R. 1864. The origin of human races and the antiquity of man deduced from the theory of "natural selection." *Jour. Anthropol. Soc.* 2:clviii–clxxvi.

Washburn, S. 1963. The study of race. *Amer. Anthropol.* 65:521–31.

Weidenreich, F. 1947. Facts and speculations concerning the origin of *Homo sapiens*. *Amer. Anthropol.* 49:187–203.

———. 1947. The trend of human evolution. *Evolution* 1:221–36.

Weiner, J. S. 1957. Physical anthropology: An appraisal. *Amer. Scientist* 45:504–509.

Witherspoon, D. J., S. Wooding, A. R. Rogers, E. E. Marchani, W. S. Watkins, M. A. Batzer, and L. B. Jorde. 2007. Genetic similarities within and between human populations. *Genetics* 176:351–59.

Wolpoff, M., and R. Caspari. 1998. *Race and human evolution: A fatal attraction*. Boulder, Colo.: Westview Press.

Wolpoff, M. H., X. Wu, and A. Thorne. 1984. Modern *Homo sapiens* origins: A general theory of hominid evolution involving the evidence from east Asia. In *The origins of modern humans: A world survey of the fossil evidence*, ed. F. H. Smith, and F. Spencer, 411–83. New York: Alan R. Liss.

CHAPTER 2. Species, Patterns, and Evolution

Crowson, R. A. 1986. *Classification and biology*. Piscataway, N.J.: Aldine Transaction.

Darwin, C. 1859. *On the origin of species by means of natural selection*. London: John Murray.

Davis, J. I., and K. C. Nixon. 1992. Populations, genetic variation, and the delimitation of phylogenetic species. *Syst. Biol.* 41:421–28.

Goldstein, P., and R. DeSalle. 2000. Phylogenetic species nested hierarchies and character fixation. *Cladistics* 16:364–84.

Gotthelf, Allan. 1985. *Aristotle on nature and living things: Philosophical and historical studies presented to David M. Balme on his seventieth birthday*. Cambridge, U.K.: Cambridge University Press.

Gould, S. J., and N. Eldredge. 1993. Punctuated equilibrium comes of age. *Nature* 366:223–27.

Grant, P. R. 1986. *The evolution and ecology of Darwin's finches*. Princeton, N.J.: Princeton University Press.

Gregg, J. R. 1954. *The language of taxonomy*. New York, Columbia University Press.

Harrison, R. G. 1991. Molecular changes at speciation. *Ann. Rev. Ecol. Syst.* 22: 281–308.

Hebert, P. D. N., A. Cywinska, S. L. Ball, and J. R. deWaard. 2003. Biological identifications through DNA barcodes. *Proc. Roy. Soc. Lond. Ser.* B:270:313–21.

Hennig, W. 1966. *Phylogenetic systematics*. English translation. Urbana: University of Illinois Press.

Kingman, J. F. C. 1982. The coalescent. *Stoch. Process. Appl.* 13:235–48.

Kreitman, M. 1983. Nucleotide polymorphism at the alcohol dehydrogenase gene region of *Drosophila melanogaster*. *Nature* 304:412–17.

Levy, S., G. Sutton, P. C. Ng, L. Feuk, A. L. Halpern, B. P. Walenz, N. Axelrod, et al. 2007. The diploid genome sequence of an individual human. *PLoS Biol*. 5: 2113–44.

Lewontin, R. C. 1974. *The genetic basis of evolutionary change*. New York: Columbia University Press.

Lipscomb, D., N. Platnick, and Q. Wheeler. 2003. The intellectual content of taxonomy: a comment on DNA taxonomy. *Trends Ecol. Evol.* 18:65–66.

Linnaeus, C. 1758. *Systema naturae. Regnum animale, Editio decima*. Facsmile ed. 1956, London: British Museum (Natural History).

Mallet, J. 2008. Hybridization, ecological races and the nature of species: Empirical evidence for the ease of speciation. *Philos. Trans. R. Soc. Lond. B. Biol. Sci.* 363:2971–86.

Marcus, G. 2008. *Kluge: The haphazard evolution of the human mind*. New York: Houghton Mifflin Harcourt.

Margulis, L., K. V. Schwartz, and M. Dolan. 1994. *The illustrated five kingdoms: A guide to the diversity of life on earth*. New York: HarperCollins College Publishers.

Mayr, E. 1942. *Systematics and the origin of species*. New York: Columbia University Press.

———. 1982. *The growth of biological thought: Diversity, evolution, and inheritance*. Cambridge, Mass.: Harvard University Press.

———. 1992. Darwin's principle of divergence. *Jour. Hist. Biol.* 25:343–59.

Medawar, P. B., and J. S. Medawar. 1983. *Aristotle to zoos: A philosophical dictionary of biology*. Cambridge, Mass.: Harvard University Press.

Ng, P.C., Q. Zhao, S. Levy, R. L. Strausberg, and J. C. Venter. 2008. Individual genomes instead of race for personalized medicine. *Clin. Pharm. Therapeut.* 84: 306–309.

Nixon, K. C., and Q. D. Wheeler. 1990. An amplification of the phylogenetic species concept. *Cladistics* 6:211–23.

Rieseberg, L. H. 1997. Hybrid origins of plant species. *Ann. Rev. Ecol. Syst.* 28: 359–89.

SA2000. *Systematics agenda 2000: Charting the biosphere*, technical report. New York: American Museum of Natural History.

Saiki, R. K., D. H. Gelfand, S. Stoffel, S. J. Scharf, R. Higuchi, G. T. Horn, K. B. Mullis, and H. A. Erlich. 1988. Primer-directed enzymatic amplification of DNA with a thermostable DNA polymerase. *Science* 239:487–91.

Simpson, G. G. 1967. *The meaning of evolution*. Rev. ed. New Haven, Conn.: Yale University Press.

Stearn, W. T. 1959. The background of Linnaeus's contributions to the nomenclature and methods of systematic biology. *Syst. Zool.* 8:4–22

Tautz, D. 1998. Debatable homologies. *Nature* 395:17–19.

Tautz, D., P. Arctander, A. Minelli, R. H. Thomas, and A. P. Vogler. 2003. A plea for DNA taxonomy. *Trends Ecol. Evol.* 18:70–74.

Templeton, A. R. 1989. The meaning of species and speciation: a genetic perspective. In *Speciation and Its Consequences*, ed. D. Otte and J. A. Endler. Sunderland, Mass.: Sinauer Associates.

Templeton A. R., E. Routman, and C. A. Phillips. 1995. Separating population structure from population history: A cladistic analysis of the geographical distribution of mitochondrial DNA haplotypes in the tiger salamander, *Ambystoma tigrinum*. *Genetics* 140:767–82.

Templeton, A. R. 2006. *Population genetics and microevolutionary theory*. Hoboken, N.J.: Wiley-Liss.

Wake, D. B. 2001. Speciation in the round. *Nature* 409:299–300.

Yoon, C. K. 2009. *Naming nature: The clash between instinct and science*. New York: Norton.

CHAPTER 3. Human Evolution and Dispersal

Bermudez de Castro, J. M., J. L. Arsuaga, E. Carbonell, A. Rosas, I. Martínez, and M. Mosquera. 1997. A hominid from the Lower Pleistocene of Atapuerca, Spain: Possible ancestor to Neandertals and modern humans. *Science* 276: 1392–95.

Bouzzougar, A., N. Barton, M. Vanhaeren, F. d'Errico, S. Collcutt, T. Highham, E. Hodge, et al. 2007. 82,000-year-old shell beads from North Africa and implications for the origins of modern human behavior. *Proc. Nat. Acad. Sci. USA* 104: 9964–69

Brown, K. S., C. W. Marean, A. I. R. Herries, Z. Jacobs, C. Tribolo, D. Braun, D. L. Roberts, M. C. Meter, and J. Bernatchez. 2009. Fire as an engineering tool of early modern humans. *Science* 325:859–62.

Brunet, M., F. Guy, D. Pilbeam, H. T. Mackaye, A. Likius, D. Ahounta, A. Beauvilain, et al. 2002. A new hominid from the Upper Miocene of Chad, Central Africa. *Nature* 418:145–51.

Cáceres, M., J. Lachuer, M. A. Zapala, J. C. Redmond, L. Kudo, D. H. Geschwind, D. J. Lockhart, et al. 2003. Elevated gene expression levels distinguish human from non-human primate brains. *Proc. Natl. Acad. Sci. USA* 100:13030–35.

Carbonell, E., J. M. Bermúdez de Castro, A. Parés, A. Pérez-González, M. Olle, M. Mosquera, G. Cuenca-Bescós, et al. 2008. The first hominin species of Europe. *Nature* 452:465–70.

Chou, H.-H., H. Takematsu, S. Diaz, J. Iber, E. Nickerson, K. L. Wright, E. A, Muchmore, et al. 1998. A mutation in human CMP-sialic acid hydroxylase occurred after the *Homo-Pan* divergence. *Proc. Natl. Acad. Sci. USA* 95:11751–56.

Dart, R. A. 1953. The predatory transition from ape to man. *Int. Anth. Linguist. Rev.* 1:201–17.

Deacon, H., and J. Deacon. 1999. *Human beginnings in South Africa: Uncovering the secrets of the Stone Age*. Cape Town: David Philip.

Ely, B., J. L. Wilson, F. Jackson, and B. A. Jackson. 2006. African-American mitochondrial DNAs often match mtDNAs found in multiple African ethnic groups. *BMC Biol.* 4:34.

Enard, W., P. Khaitovich, J. Klose, S. Zöllner, F. Heissig, P. Giavalisco, K. Nieselt-Struwe, et al. 2002. Intra- and interspecific variation in primate gene expression patterns. *Science* 296:340–43.

Gabunia, L., A. Vekua, D. Lordkipanidze, C. C. Swisher, R. Ferring, A. Justus, M. Nioradze, et al. 2000. Earliest Pleistocene hominid cranial remains from Dmanisi, Republic of Georgia: Taxonomy, geological setting and age. *Science* 288: 1019–25.

Gilad, Y., A. Oshlack, G. K. Smyth, T. P. Speed, and K. P. White. 2006. Expression profiling in primates reveals a rapid evolution of human transcription factors. *Nature* 440:242–45.

Hart, D., and R. W. Sussman. 2005. *Man the hunted.* Boulder, Colo.: Westview Press.

Henshilwood, C., F. d'Errico, R. Yates, Z. Jacobs, C. Tribolo, G. A. T. Duller, N. Mercier, J. C. Sealy, H. Valladas, I. Watts, and A. G. Wintle. 2003. Emergence of modern human behavior: Middle Stone Age engravings from South Africa. *Science* 295:1278–80.

Henshilwood, C., F. d'Errico, M. Vanhaeren, K. van Niekerk, and Z. Jacobs. 2004. Middle Stone Age shell beads from South Africa. *Science* 304:404.

Khaitovich, P., B. Muetzel, X. She, M. Lachmann, I. Hellmann, J. Dietzsch, S. Steigele, et al. 2004. Regional patterns of gene expression in human and chimpanzee brains. *Genome Research* 14 (8):1462–73.

King, M. C., and A. C. Wilson. 1975. Evolution at two levels in humans and chimpanzees. *Science* 188:107–16.

Klein, R. G. 1999. *The human career.* 2d ed. Chicago: University of Chicago Press.

Lordkipanidze, D., T. Jashashvili, A. Vekua, M. S. Ponce de Leon, C. P. E. Zollikofer, G. P. Rightmire, H. Pontzer, et al. 2007. Postcranial evidence from early *Homo* from Dmanisi, Georgia. *Nature* 449:305–10.

McDougall, I., F. H. Brown, and J. G. Fleagle. 2005. Stratigraphic placement and age of modern humans from Kibish, Ethiopia. *Nature* 433:733–36.

Marean, C. W., M. Bar-Matthews, J. Bernatchez, E. Fisher, P. Goldberg, A. I. R. Herries, Z. Jacobs, et al. 2008. Early human use of marine resources and pigment in South Africa in the Middle Pleistocene. *Nature* 449:905–909.

Novembre, J., T. Johnson, K. Bryce, Z. Kutalik, A. R. Boyko, et al. 2008. Genes mirror geography within Europe. *Nature* 456:98–101.

Panger, M., and C. Boesch. 2002. Excavation of a chimpanzee stone tool site in the African rainforest. *Science* 296:1452–55.

Pickford, M., and B. Senut. 2001. The geological and faunal context of Late Miocene hominid remains from Lukeino, Kenya. *Earth Planet. Sci.* 332:145–52.

Povinelli, D. J. 2004. Behind the ape's appearance: Escaping anthropocentrism in the study of other minds. *Daedalus* 133 (1):29–41.

Rogers, A. R., D. Iltis, and S. Wooding. 2004. Genetic variation at the MC1R locus and the time since loss of human body hair. *Curr. Anthropol.* 45:105–108.

Schick, K. D., and N. Toth. 1993. *Making silent stones speak: Human evolution and the dawn of technology.* New York: Simon and Schuster.

Schick, K. D., N. Toth, G. Garufi, S. Savage-Rumbaugh, D. Rumbaugh, and R. Sevcik. 1999. Continuing investigations into the stone tool-making and tool-using capabilities of a bonobo (*Pan paniscus*). *Jour. Archaeol. Sci.* 26:821–32.

Shriver, M. D., G. C. Kennedy, E. J. Parra, H. A. Lawson, V. Sonpar, J. Huang, J. M. Akey, and K. W. Jones. 2004. The genomic distribution of population substructure in four populations using 8,525 autosomal SNPs. *Human Genomics* 1: 274–86.

Swisher, C. C., III, W. J. Rink, S. C. Antón, H. P. Schwarcz, G. H. Curtis, A. Suprijo, and Widiasmoro. 1996. Latest *Homo erectus* of Java: Potential contemporaneity with *Homo sapiens* in Southeast Asia. *Science* 274:1870–74.

Swisher, C. C., III, G. H. Curtis, T. Jacob, A. G. Getty, A. Suprijo, and Widiasmoro. 1994. Age of the earliest known hominids in Java, Indonesia. *Science* 263: 1118–21.

Tattersall, I. 2008. An evolutionary framework for the acquisition of symbolic cognition by *Homo sapiens*. *Comp. Cogn. Behav. Revs.* 3:99–114.

———. 2009. *The fossil trail: How we know what we think we know about human evolution.* 2d ed. New York: Oxford University Press.

Tattersall, I., and J. H. Schwartz. 2006. The distinctiveness and systematic context of *Homo neanderthalensis*. In *Neanderthals revisited: New approaches and perspectives*, ed. K. Harvati and T. Harrison, 9–22. New York: Springer.

———. 2009. Evolution of the genus *Homo*. *Ann. Rev Earth Planet. Sci.* 37:67–92.

Templeton, A. R. 2005. Haplotype trees and modern human origins. *Yrbk Phys. Anthropol.* 48:33–59.

Thieme, H. 1997. Lower Palaeolithic hunting spears from Germany. *Nature* 385: 807–10.

Tishkoff, S. A., F. A. Reed, F. R. Friedlaender, C. Ehret, A. Ranciaro, A. Froment, J. B. Hirbo, et al. 2009. The genetic structure and history of Africans and African Americans. *Science* 324:1035–44.

Uddin, M., D. E. Wildman, G. Liu, W. Xu, R. M. Johnson, P. R. Hoff, G. Kapatos, et al. 2004. Sister grouping of chimpanzees and humans as revealed by genome-wide phylogenetic analysis of brain-gene expression profiles. *Proc. Natl. Acad. Sci. USA* 101:2957–62.

Valladas, H., J. L. Reyss, J. L. Joron, G. Valladas, O. Bar-Yosef, and B. Vandermeersch. 1988. Thermoluminescence dating of Mousterian "Proto-Cro-Magnon" remains from Israel and the origin of modern man. *Nature* 331:614–16.

Vanhaeren, M., F. d'Errico, C. Stringer, S. L. James, J. A. Todd, and H. K. Mienis. 2006. Middle Paleolithic shell beads in Israel and Algeria. *Science* 312:1785–88.

Vekua, A., D. Lordkipanidze, G. P. Rightmire, J. Agusti, R. Ferring, G. Majsuradze, A. Mouskhelishvili, M. Nioradze, M. Ponce de Leon, M. Tappen, M. Tvalchrelidze, and C. Zollikofer. 2002. A new skull of early *Homo* from Dmanisi, Georgia. *Science* 297:85–89.

White, T. D., B. Asfaw, D. DeGusta, H. Gilbert, G. D. Richards, G. Suwa, and F. C. Howell. 2003. Pleistocene *Homo sapiens* from Middle Awash, Ethiopia. *Nature* 423:742–47.

White, T. D., organizer. 2009. Special issue on *Ardipithecus ramidus*. *Science* 326:5–106.

Zollikofer, C. P. E., M. S. Ponce de León, D. E. Lieberman, F. Guy, D. Pilbeam, A. Likius, H. T. Mackaye, P. Vignaud, and M. Brunet. 2005. Virtual cranial reconstruction of *Sahelanthropus tchadensis*. *Nature* 434:755–59.

Zuckerkandl, E., and L. Pauling. 1965. Evolutionary divergence and convergence in proteins. In *Evolving genes and proteins*, ed. V. Bryson and H. J. Vogel, 97–166. New York: Academic Press.

CHAPTER 4. Is "Race" a Biological Problem?

Balter, M. 2005. Zebrafish researchers hook gene for human skin color. *Science* 310:1754–55.

Barsh, G. S. 2003. What controls variation in human skin color? *PLoS Biol.* 1 (1):e27.

Beall, C. M. 1981. Optimal birthweights in Peruvian populations at high and low altitudes. *Amer. Jour. Phys. Anthropol.* 56:209–16.

———. 1982. A comparison of chest morphology in high altitude Asian and Andean populations. *Hum. Biol.* 54:145–63.

———. 2003. High altitude adaptations. *The Lancet* 362:S14–S15.

Beja-Pereira, A., G. Luikart, P. R. England, D. G. Bradley, O. C. Jann, G. Bertorelle, A. T. Chamberlain, et al. 2003. Gene-culture coevolution between cattle milk protein genes and human lactase genes. *Nature Genetics* 35:311–13.

Bersaglieri, T., P. C. Sabeti, N. Patterson, T. Vanderploeg, S. F. Schaffner, J. A. Drake, M. Rhodes, D. R. Reich, and J. N. Hirschhorn. 2004. Genetic signatures of strong recent positive selection at the lactase gene. *Amer. Jour. Hum. Genet.* 74: 1111–20.

Churchill, S. E., Shackelford, L. L., J. N. Georgi, and M. T. Black. 2004. Morphological variation and airflow dynamics in the human nose. *Amer. Jour. Hum. Biol.* 16: 625–38.

Freeman, R. B. 1984. *Darwin pedigrees*. London: privately printed.

Hollox, E. 2005. Genetics of lactase persistence—fresh lessons in the history of milk drinking. *Eur. Jour. Hum. Genet.* 13:267–69.

Jablonski, N. 2006. *Skin: A natural history*. Berkeley: University of California Press.

Lalueza-Fox, C., H. Römpler, D. Caramelli, C. Stäubert, G. Catalano, D. Huges, N. Rohland, et al. 2007. A melanocortin 1 receptor allele suggests varying pigmentation among Neanderthals. *Science* 318:1453–55.

Lamason, R. L., M. P. K. Mohideen, J. R. Mest, A. C. Wong, H. L. Norton., M. C. Aros, M. J. Jurynec, et al. 2005. SLC24A5, a putative cation exchanger, affects pigmentation in zebrafish and humans. *Science* 310:1782–86.

Lee, S., and S. J. Piazza. 2009. Built for speed: Musculoskeletal structure and sporting ability. *Jour. Exper. Biol.* 212:3700–3707.

Moore, L. G., S. Niermeyer, and S. Zamudio. 1998. Human adaptation to high altitude: Regional and life-cycle perspectives. *Yrbk Phys. Anthropol.* 41:25–64.

Rogers A., D. Iltis, and S. Wooding. 2004. Genetic variation at the MC1R locus and the time since loss of human body hair. *Curr. Anthropol.* 45:105–108.

Swallow, D. M. 2003. Genetics of lactase persistence and lactase intolerance. *Ann. Rev. Genet.* 37:197–219.

CHAPTER 5. Race in Ancestry, Forensics, and Disease

Bolnick, D. A. 2008. Individual ancestry inference and the reification of race as a biological phenomenon. In *Revisiting race in a genomic age*, ed. B. A. Koenig, S. S.-J. Lee, S. S. Richardson, 89–101. New Brunswick, N.J.: Rutgers University Press.

Bolnick, D., D. Fullwiley, T. Duster, R. S. Cooper, J. H. Fujimura, J. Kahn, J. S. Kaufman, J. Marks, A. Morning, A. Nelson, P. Ossorio, J. Reardon, S. Reverby, and K. TallBear. 2007. The science and business of genetic ancestry testing. *Science* 318:399–400.

Cheung, V. G., R. S. Spielman, K. T. Ewens, T. M. Weber, M. Morley, and J. T. Burdick. 2005. Mapping determinants of human gene expression by regional and genome-wide association. *Nature* 437:1365–69.

Gibbs, R. 2005. Deeper into the genome. *Nature* 437:1233–34.

Goldstein, D. B., K. R. Ahmadi, M. E. Weale, and N. W. Wood. 2003. Genome scans and candidate gene approaches in the study of common diseases and variable drug responses. *Trends Genet.* 19:615–22.

Goldstein, D. B., and G. L. Cavalleri. 2005. Understanding human diversity. *Nature* 437:1241–42.

International HapMap Consortium. 2005. A haplotype map of the human genome. *Nature* 437:1299–1320.

———. 2007. A second generation human haplotype map of over 3.1 million SNPs. *Nature* 449:851–61.

International HapMap Project. http://snp.cshl.org/

Jakobsson, M., M. Jakobsson, S. W. Scholz, P. Scheet, J. R. Gibbs, J. M. VanLiere, Hon-C. Fung, Z. A. Szpiech, et al. 2008. Genotype, haplotype and copy-number variation in worldwide human populations. *Nature* 451:998–1003.

Kahn, J. 2006. Race, pharmacogenomics, and marketing: putting BiDil in context. *Am. J. Bioeth.* 6:W1–5.

Kim, J. I., J.-I. Kim, Y. S. Ju, H. Park, S. Kim, S. Lee, J.-H. Yi, J. Mudge, et al. 2009.

A highly annotated whole-genome sequence of a Korean individual. *Nature* 460:1011–15.

Koenig, B. A., S. S.-J. Lee, S. S. Richardson, eds. 2008. *Revisiting race in a genomic age.* New Brunswick, N.J.: Rutgers University Press.

Lander, E. S., and D. Botstein. 1989. Mapping mendelian factors underlying quantitative traits using RFLP linkage maps. *Genetics* 121:185–99.

Lee, S. S.-J., D. A. Bolnick, T. R. Duster, P. Ossorio, and K. TallBear. 2009. The illusive gold standard in genetic ancestry testing. *Science* 325:38–39.

Lehrman, S. From race to DNA. *Scientific American,* February 2008.

Lupski, J. R., J. G. Reid, C. Gonzaga-Jauregui, D. R. Deiros, D. C. Y. Chen, L. Nazareth, M. Bainbridge, et al. 2010. Whole genome sequencing of a patient with Charcot-Marie-tooth neuropathy. *New England Journal of Medicine* 362: 1181–91.

McCarthy, M. I., G. R. Abecasis, L. R. Cardon, D. B. Goldstein, J. Little, J. P. A. Ioannidis, and J. N. Hirschhorn. 2008. Genome-wide association studies for complex traits: consensus, uncertainty and challenges. *Nat. Rev. Genet.* 9:356–69.

McKernan, K. J., H. E. Peckham, G. L. Costa, S. F. McLaughlin,Y. Fu1, E. F. Tsung, C. R. Clouser, et al. 2009. Sequence and structural variation in a human genome uncovered by short-read, massively parallel ligation sequencing using two-base encoding. *Genome Res.* 19:1527–41.

Moore, J. 2005. Breeding: Darwin doubted his own family's "fitness." *Natural History,* November 2005: 45–46.

Need, A., and D. B. Goldstein. 2009. Next generation disparities in human genomics: Concerns and remedies. *Trends Genet.* 25:489–90.

Ng, P. C., Q. Zhao, S. Levy, R. L. Strausberg, and J. C. Venter. 2008. Individual genomes instead of race for personalized medicine. *Clinical Pharmacology & Therapeutics* 84:306–309.

Parra, E. J., R. A. Kittles, G. Argyropoulos, C. L. Pfaff, K. Hiester, C. Bonilla, N. Sylvester, et al. 2001. Ancestral proportions and admixture dynamics in geographically defined African Americans living in South Carolina. *Amer. Jour. Phys. Anth.* 114:18–29.

Perkins, H. F. 1934. *A decade of progress in eugenics: Scientific papers of the Third International Congress of Eugenics.* Baltimore: Williams and Wilkins Company.

Reich, D. E., and E. S. Lander. 2001. On the allelic spectrum of human disease. *Trends Genet.* 17:502.

Risch, N., and K. Merikangas. 1996. The future of genetic studies of complex human diseases. *Science* 273:1516–17.

Roach, J. C., G. Glusman, A. F. A. Smit, C. D. Huff, R. Hubley, P. T. Shannon, L. Rowen, et al. 2010. Analysis of genetic inheritance in a family quartet by whole-genome sequencing. *Science* 328:636–39.

Schubert, C. 2008. Interview with Charles Rotimi. *Nature Medicine* 14:704–705.

Schuster, S. C., W. Miller, A. Ratan, L. P. Tomsho, B. Giardine, L. R. Kasson, R. S. Harris, et al. 2010. Complete Khoisan and Bantu genomes from southern Africa. *Nature* 463:943–47.

Simón-Sánchez, J., and A. Singleton. 2008. Genome-wide association studies in neurological disorders. *Lancet Neurol.* 7:1067–72.

Slatkin, M. 1999. Disequilibrium Mapping of a Quantitative-Trait Locus in an Expanding Population. *Amer. Jour. Hum. Genet.* 64:1765–73.

———. 2008. Linkage disequilibrium—understanding the evolutionary past and mapping the medical future. *Nat. Rev. Genet.* 9:477–85.

Travis, J. 2009. Scientists decry isotope, DNA testing of "nationality." *Science* 326:30–31.

Wang, J., W. Wang, R. Li, Y. Li, G. Tian, L. Goodman, W. Fan et al. 2008. The diploid genome sequence of an Asian individual. *Nature* 456:60–65.

Wheeler, D. A., M. Srinivasan, M. Egholm, Y. Shen, L. Chen, A. McGuire, W. He, et al. 2008. The complete genome of an individual by massively parallel DNA sequencing. *Nature* 452:872–76.

EPILOGUE

Relethford, J. H. 2009. Race and global patterns of phenotypic variation. *Amer. Jour. Phys. Anthropol.* 139:16–22.

INDEX

Letters following a page number denote: figures (f), tables (t)

OTHER TITLES IN THE

TEXAS A&M UNIVERSITY ANTHROPOLOGY SERIES: